Architectural Material 3

GLASS

지은이	담디 편집부 엮음
펴낸이	서경원
편집	나진연
디자인	이철주
펴낸곳	도서출판 담디
등록일	2002년 9월 16일
등록번호	제 9-00102호
주소	서울시 강북구 삼각산로 79, 2층
전화	02-900-0652
팩스	02-900-0657
이메일	damdi_book@naver.com
홈페이지	www.damdi.co.kr

Compiler	DAMDI Publishing House
Publisher	Kyongwon Suh
Editor	Jinyoun Na
Art Director	Cheolju Lee
Publishing	DAMDI Publishing House
Office Address	2F, 79, Samgaksan-ro, Gangbuk-gu, Seoul, 01036, Korea
Tel	+82-2-900-0652
Fax	+82-2-900-0657
E-mail	damdi_book@naver.com
Homepage	www.damdi.co.kr

정가 88,000원
Printed in Korea
ISBN 978-89-6801-082-8 (94540)
978-89-6801-079-8 (set)

이 도서의 국립중앙도서관 출판예정도서목록(CIP)은 서지정보유통지원시스템 홈페이지(http://seoji.nl.go.kr)와 국가자료공동목록시스템(http://www.nl.go.kr/kolisnet)에서 이용하실 수 있습니다.(CIP제어번호: CIP2018024918)

Architectural Material 3

GLASS

담디
DAMDI

Contents

Multi - Housing

Education

Commercial

Office

Installation

Public (Urban & Landscape)

How use the Glass

Interview with Architects

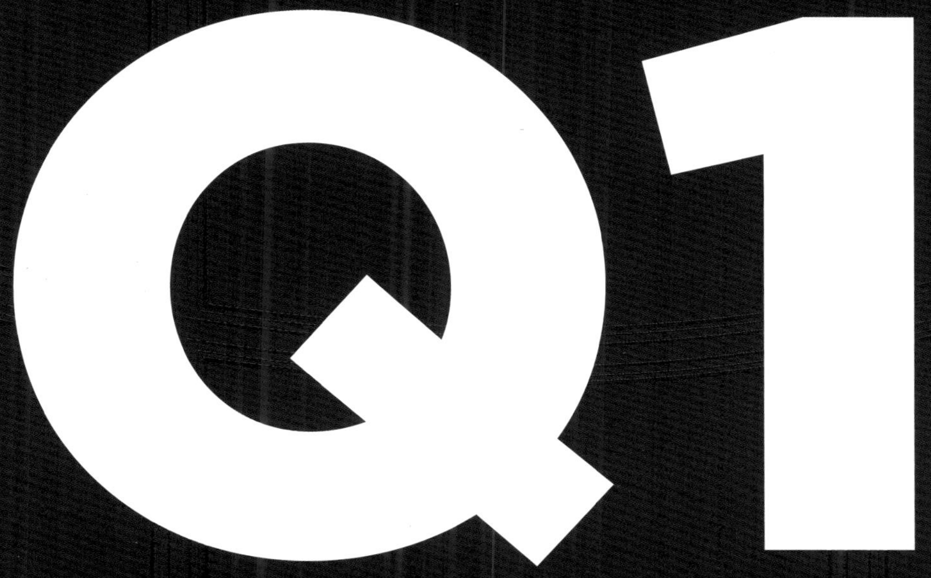

Tell us about your favourite project that you used Glass in or another architect's work

- it can be in the interior, on the facade, doesn't matter where it's used.

ARPHENOTYPE

As a design element of contemporary architecture, the building material glass is almost indispensable. The continuous development of manufacturing and processing technology and the constant improvement of ready-to-use products play a crucial role. Except for a special architecture, like military bunker, glass is used in all buildings. One famous building is the glass house by the American architect Philip Johnson(***A1***), who wanted a house made entirely of glass and consisting of only one large room. In this way he wanted to break the boundaries between inside and outside. This characteristic leads me to a new architecture based on glass (fibers). The World Wide Web. I think most beings do not see this as a building, but it is a built structure that determines our way of life. Like the glasshouse by Johnson, it breaks the boundaries between inside and outside, which is in conclusion often then controlled by governments. But like a sewage system, which connects all buildings in a city, the WWW connects through Fibre-to-the-Home (FTTH) all buildings to a global network-city; hence it is architecture and we need to develop a new consciousness, of how we, as architects plan and design in such a city. A city, which is based on glass (and data).

AZC

We used glass on a medical house project "NDSBS" in Paris and on an office building "Pacemar(**344p**)" in Suresnes. For NDBS project, glass is a cladding material. For Pacemar, glass is a ventilated double façade material.

BOARD

We recently reached the final phase in an architecture competition for the extension of the District Office in Görlitz, Germany with a project that we called "High and Dry(***A2***, **322p**)" and in which the use of glass was crucial for its success. Because in order to ensure easy orientation for guests and employees within the entire building complex we created a glazed connecting first floor, providing access to all other buildings of the administrative campus barrier-free, using one of the greatest qualities and strengths of glass: being optically transparent, yet supplying a sufficient division of the inside from the outside of a building. Through this transparent connecting first floor that we created in an elongated administration building inside the block with three wings and three bridges, all the future departments of the enlarged district office will be accessible from the entrance "high and dry".

Carlos Lampreia

A good example of using glass as a transparent wall is the deutsch pavillion(***A3***) in barcelona by Mies van der Rohe, where we can feel the abilitys of the material as a filter towards the exterior, wherever is a square or nature like in the patio. In both situation glass has the ability to provide a very strong feeling of continuity but also a very cosy internal space.

Casanova+Hernandez Architects

For the Albanian National Museum(***A4***) of Photography, Marubi, we have designed a curtain wall to cover the three facades of the

A1 Glasshouse - Philip Johnson ©Staib

A2 High and Dry

A3 Deutsch Pavilion - Mies van der Rohe ©Alice Wiegand

A4 Albanian National Museum, MARUBI ©Christian Richters

central courtyard of the building that has helped to define its strong identity.

Thanks to a specially designed structure, the partitions of the glazed facades follow a rhythmic pattern where the structure is embedded and remains hidden to the eye.

In that way, the glazed facades seem to be free from structural requirements and their pattern can expressively be used to create a rich dialogue not only among the glazed facades themselves but also between them and the rest of the building.

The courtyard becomes a 'musical' architectural element, bathed by the changing natural light captured by the glazed facades.

At night, through the transparency of the glazed facades, lights from inside the building bathe the space of the courtyard and create an intimate place for relaxing and for enjoying the decaying light of the sunset.

A5 Swiss House IV, Muzzano

CEBRA(Mikkel Frost)

Apart from the glass, you'll find in the windows of our projects, we do not use the material a lot.

I remember visiting Maison de Verre in Paris as a student some 25 years ago. It made a great impression on me. The house is designed by Pierre Chareau in 1928 and it was decades ahead of time in terms of transparency effects, details and aesthetics. I believe it is open to the public today and it is certainly worth a visit.

Davide Macullo Architects

Glass is a natural material. Glass represents first of all a transparence between inside and outside, secondly, a potential of using it for its reflection characteristics. An interesting recent project in glass is the work of Jean Nouvel for the entrance wall of the Foundation Cartier in Paris for its dimension and relation between two spaces that are both outdoor spaces but in two different conditions; street-garden. In our house in Muzzano(***A5***, **094p**) we used the same principle in a reduced intimate space where the glass has been used for its reflecting characteristics. Both are inspired by the work of the artist Dan Graham(***A6***).

A6 Pavillon - Dan Graham ©wikipedia commons

Donner Sorcinelli Architecture

The use of glass always represents an exciting challenge to be played in between transparencies and reflections. One of the project where the combined use of glass and other materials has been developed according to these aspects it was the design of a Primary School in Carbonera (Italy). In this case, the visual connection between the common spaces and classrooms with the landscape has been the main driver of the whole design process.

Katsutoshi Sasaki+Associates

National Gallery in Germany by Mies van der Rohe(***A7***).

Keiichi Hayashi Architect

Clean Room / Yuji Takeoka

Aside context in art, I like this artwork, because it simultaneously creates two spaces such as an empty space inside glass box and huddled imbrication space of spectators and other exhibitions, which

A7 new National Gallery - Mies van der Rohe ©Jean-Pierre Dalbéra

created by reflection of glass exterior.

A8 Deutcher Werkbund Pavilion Exterior

LANDÍNEZ+REY arquitectos

Undoubtedly our references for glass architecture are those of the history of architecture. In the first place, the radical and wise use of glass in the Maison de Verre by Pierre Chareau or the Pavilion by Bruno Taut for the Deutcher Werkbund(***A8***) in 1914: these are architectural lessons for glass use. In advance, the architecture of Mies or, in 1949, the Glass House of Philip Johnson. that seem to evolve towards the use of glass, today, by Kazuyo Sejima.

Our Train Station in Rivas-Futura(**396p**) (Madrid, Spain) is definitely a glass box arranged on a podium and protected with a shade covering floating above it, according to our climate. By linear aluminum louvers in front of the glazing, a bioclimatic shade covering is created.

It offers protection against west solar direct radiation avoiding overheating and provides reflected natural lighting inside the spaces avoiding glare. Also, a covered exterior area where the external stairs between public spaces is provided.

m artı d mimarlık

Izmir Chamber(**050p**) of Geological Engineers is a small scale project that we have designed on a narrow parcel. It is located on an attached parcel. Services, circulation elements are located at the back side of the parcel which enables building to get light in from one facade of it. For maximising use of natural light in the building, glass curtain wall was used on that facade. In order to, reflect institutional identity, topographicalreferences are used for geometry of the facade.

modostudio

An architectural icon that was never build is the Mies van der Rohe Glass Skyscraper (1922). Mies van der Rohe was one of the first architect to understand the poetics of glass and to bring this material to extreme solutions.

Mork-Ulnes Architects

It is hard not to think of Mies – with either the Farnsworth house or the Friedrichstrasse skyscraper project as being the most groundbreaking use of glass in the modernist sense.

A9 Refurbished an Office, ©Pedro Nuno Pacheco

murmuro

We have refurbished an office(***A9***, **316p**), actually, transformed three small office spaces into a bigger one. In this project we have used glass as a partition material that would allow us to keep the flow and open feel needed. The main workroom, for about 40 people, has an entire side in glass/window frames for the outside, and the two smaller sides also in glass, facing the kitchen and lounge area on one side and the partners office/meeting room on the other.

NISHIZAWA ARCHITECTS

BEN THANH restaurant(**068p**) is our favourite project that using glass for the partition inside.

object-e architecture

Out project for the Cyprus Medical School competition(***A10, Next Page***) is based on the principle that a medical school, as any other university building, shouldn't be just a place where the students go in order to attend classes and then leave. Instead the proposal tries to imagine the design proposal as a place where the students, along with instructors and administration stuff, spend an important part of their day and should therefore offer them high quality spaces not only for their main activity, but also in the spaces intended for leisure, meeting and rest.

In that direction the proposal echoes a cloister organization: It expands to all four edges of the site allowing for the creation of a large open space in the center of the building. This space, organized in levels and planted extensively, is understood as the heart of the building. That interior courtyard, while large in size and the main organizational element of the building, remains hidden from the outside. Aim was to function as a secret space; a surprise for the visitor that approaches the building. The exterior of the building on the other hand becomes 'monolithic': it leaves only a couple of small openings through which you can see the interior courtyard while some vegetation that appears on the elevations provides some hints for what exists on the inside.

Classrooms, labs and offices are placed on the outside perimeter of the building, while the interior is occupied by a large corridor facing the courtyard through transparent glass panels. The interior elevations therefore – the sides of the building that face the atrium – are transparent: A glass elevation opens from the building towards the courtyard allowing a constant visual connection between the atrium and the inside of the building. The glass is structured in a rhythmical way though mullions that vary, generating a non-repetitive pattern. The transparency of the glass is controlled only through vegetation that rises in front of the glass panels in places that full transparency in not required.

A10 the Cyprus Medical School competition

OFIS arhitekti

So far, the most interesting project we have used glass for is the Glass House in the desert in Spain. The unit uses the vertical glazing panels of the envelope as structural walls, resisting the desert's high-speed winds and supporting the timber stressed skin roof and deck. The glass thermally efficient envelope is constructed of triple glazing walls that, due to use of almost invisible coatings, protects the interior from the sun.

OOIIO Architecture

For us, glass never has been the main material of any of our projects. Of course we used glass in every building that we did, in windows, dividing walls, etc. but for the moment never as the main material that define a project.

I am sure it is because until now I have been working on regions were the sun heat is really strong especially in summertime, like central Spain, were we have our office.

Glass is not the most appropriate material over there. Also the local architectural tradition is based more in opaque and massive materials than in light clear ones like glass.

Piuarch

Glass with different inclinations, coloured, with specific films for the reflection of the sun, or curved; with very large panels or in thin sheets. Our projects are a real manual for the use of this material, widely adopted to allow the constant quality of natural lighting within the environments.

The Quattro Corti project, in the heart of Saint Petersburg, is certainly

A11 project Pisal

A12 Science Island and Innovation Center in Lithuania

A13 Science Island and Innovation Center in Lithuania

A14 Maison de Verre, © SEIER + SEIER

worth to mention: the historical facades, which are subject to restrictions, of the two pre-existing buildings conceal four courtyards characterised by four different colours: red, green, grey and blue define the façades of each of the courtyards, made with glass elements placed at different angles that generate a kaleidoscopic effect of great effect.

The Ekaterinensky Congress Centre in Krasnodar is also in the executive phase, offering an experimental and innovative use of glass: the front facing the river 'Kuban consists of curved frames that, together with the special treatment of glass, reflects the ripples of water.

SLOT STUDIO

In our project Pisal(***A11***) even though glass was the least used material it became the main character of it all. A massive concrete structure works around a glass viewer specially designed to overlook an artistic floor. Like a habitable kaleidoscope.

SMAR Architecture Studio

SMAR's current project for "Science Island and Innovation Center in Lithuania"(***A12, 13***), currently to be completed in 2020 it is a project that celebrates transparency. The 300m long glass wall, with Low iron super white panels of 8x1.60m creates a façade that dialogues with nature. On the other hand the 3 courtyards of 27m of diameter of glass separates the spaces in the building, give visitors visual contact with the outside, the Museum activities, and exhibition spaces, at all times. One of the most interesting things in this project is the way it creates an intimate relationship between the inside and the outside, giving visitors the feeling of walking under the sky...feeling the green atmosphere of the island.

Stefano Corbo STUDIO

The Maison de Verre (House of Glass, ***A14***) is a project designed by Pierre Chareau in Paris, from 1928 to 1932. Glass bricks are applied to the façade of the building as a homogeneous translucent wall, with some specific openings to allow transparency and air circulation.

At night, when artificial illumination is on, the house acts as a lighthouse, and is perceived as a dematerialized ephemeral object.

stpmj Architecture

Glass Farm/MVRDV, the original materiality of glass, transparency transformed to a new materiality of glass by printing the images on the glass.

Studio Farris Architects

We used glass for the Flemish parliament pavilion, for the Park tower in Antwerp, for the City Library in Bruges.

SUPA architects schweitzer song

Glass we would almost not consider a "material" in architecture but an unavoidable necessity of architecture. Again, for us glass is most fascinating when it is used structurally like in experimental pavilions or frameless just as itself.

TAKK Architecture

The glass pavilion at the Toledo Museum(***A15, Next Page***) of Art by SANAA. We find that it is a building that, by just using glass it makes us feel all the properties of the material.

TOUCH Architect

Café Pixel(***A16***) is located at the city center of Udonthani, Thailand. It has a limited land area which can be approached from main street directly. There are three primary functions which are café, restaurant, and bar, together in one space with time-sharing functions; café / bakery and healthy food restaurant in the morning, while using as a craft-beer bar in the evening.

Since it is a time-sharing, massing and zoning design of this café was created in the way to suit these both contrast functions. There are two floors, the first floor with café and bar counter, and the mezzanine floor for only customers' seats, which fit in a rectangular function box, by concerning building codes, rules, and regulations. In order to maximize the amount of seats, an open-plan and flexible seating is needed.

One simple box was turning to a pixelate box with 1x1 meter of each pixel. It creates a façade pattern, continuing into inside space. The pixel is not only used for façade, but also integrated and used as 'pixel bar seats' and a decoration shelf for displaying beverages items.

Material of the pixel is all glass, which contains both transparent clear glass and translucent glass, in order to portray different effects through each type of material and each amount of glass layers; single or double. White translucent glass will help reduce and reflect direct sunlight during daytime for café, while lighting design inside will glow as a craft-beer bar during nighttime. It creates different atmosphere for different time.

UNStudio

In the La Defense offices in Almere (NL) the facades are clad with glass panels in which a multi-coloured foil is integrated and, depending on the time of day and the angle at which you view them, a variety of different colours are reflected. The office workers there have said that the effect is as if the sun shines in that courtyard every day of the year.

A15 Toledo - Glass Pavillion ©Adam C Nelson

A16 Café Pixel

What are the strengths and weaknesses of Glass?

ARPHENOTYPE

The raw materials for glass are practically unlimited available. However, it is relatively sensitive to mechanical stress, if not specific made. Probably a downside of glass is its cost, especially when it comes to special glass constructions.

AZC

The strengths are the transparence, the lightness, the long-lasting with no maintenance.

Weakness, the price and the fact that in some extreme temperature areas, hot and cold it does not protect the interior.

BOARD

When it comes to architecture, one of the considerable weaknesses of glass - next to its porousness that exposes it to the constant risk of cracking – is its ability to transmit light and heat, if too thin and unprotected, that can easily overheat interior spaces during sunny days. To avoid that from happening we proposed to install aluminium profiles as Brise Soleil with a shading effect all over the glass facade of the "High and Dry" project. The use of triple glazing helped keeping the interior spaces in moderate temperatures during the summer months. The proposed highly effective sunscreen made of stainless steel limbweave also ensured agreeable interior temperatures during hot days. All these measures led to a highly energy efficient administrative building.

Carlos Lampreia

Glass is a marveilous material, because in strict sense we can get a transparent wall using it. But sometimes we commit the error of looking at glass as an invisible material, and thats a mistake. Glass its the most visible of all materials, because it is a perfect and shining higlhy procesed material. The bigger weakness of glass is its fragility and dependence on other subtructers that are required while using it.

Casanova+Hernandez Architects

As mentioned before, for us the potentials of using glass are enormous.

Glasses can be transparent or can present a wide range of translucency, they can be colored, can be printed, can create rich reflections or can be matt.

Nowadays it is possible for the designer to express very different architectonic intentions by using different glass qualities, by combining glasses of different properties, by curving them or by creating contrast with other materials that will be reflected on them. Glass is the most versatile material in our opinion. With its properties, it translates poetical intentions into architecture in a unique way.

CEBRA(Mikkel Frost)

The building codes in Denmark are rather strict when it comes to energy efficiency. So, when we design big glazed window areas, we struggle with heat loss in the winter and overheating in the summer. Glass as a cladding material is quite splendid though. We used it for our great mirror space in The Grundfos Halls of Residence(**190p**). It´s funny though, glass is probably the only material that an architect cannot totally avoid. It would be like cooking without salt...

Davide Macullo Architects

Glass is the material of wonder: it can be the most solid or the most inexistent material, the most challenging and the most dangerous material. Too much reflection and transparence is against human nature. The extensive use of glass in building industry has brought the rise of a series of uncomfortable situations where people have increased their level of stress. I refer to situations such as glass buildings in metropolitan areas, for example London office spaces suffer from this extended use of glass for the working habitat of humans.

Donner Sorcinelli Architecture

There are no strengths or weaknesses but poetics' potentials only.

Katsutoshi Sasaki+Associates

Weakness point is Heat insulating performance.

There are a lot of good points.

Keiichi Hayashi Architect

Strength: It is transparent material.

Weakness: It is too strong and too suddenly united between inside and outside through glass without consciousness.

LANDÍNEZ+REY arquitectos

The strengths and weaknesses of the use of glass in architecture have been linked to their intrinsic values around transparency, or not, or to their light-responsive properties.

Today the classic discourse of the use of glass in architecture has been added in an essential way its response in relation to the energetic factors.

m artı d mimarlık

Glass is advantegous due to its transparent characterisatics. However, bad heat control and sound insulation characteristics are improved by technological developments. Layerd use of glass with air gaps and coating it by various materials, improves its heat insulation capacity. As a result use of glass becomes more easier and widespread.

modostudio

The strengths of the glass is its transparency, a characteristic which can become a weakness too. Glass is a material that has to be used with extreme care. It can bring lightness but only if carefully technological detailed.

Mork-Ulnes Architects

Glass opens up so many opportunities to allow for more open, light filled structures.

A weakness of glass for us can be when we use too much glazing. It is often a temptation to generate too much transparency - taking away the power to frame views and manipulate light.

murmuro

Glass has the capability of defining a physic barrier and at the same time, not. It´s weakness may be that is difficult to achieve acoustic comfort with it.

NISHIZAWA ARCHITECTS

Strength: Can help to create an in-between space and also connect the space inside and outside together.

With its transparent and reflection character, when the sunlight reflect into the glass, the angle of it can help to create the variaty of interesting optical effects with surrounding context at different points of time, weathers or seasons. We think that it's really make sense when that partition can bring a refresh feeling for everyone, everytime they come to our building design.

Weakness: Sometimes still need some frames to protect the edges and surfaces. For this case, those glass was like a hand-craft so it's really take time and money to make them reality to our building.

Not totally stop the air conditioner escaping from the inside space.

Need cleaning up regularly cause it easily get dirty

object-e architecture

"The longing for purity and clarity, for glowing lightness and crystalline exactness, for immaterial lightness and infinite liveliness found a means of its fulfillment in glass—the most ineffable, most elementary, most flexible and most changeable of materials, richest in meaning and inspiration, fusing with the world like no other. This least fixed of materials transforms itself with every change of atmosphere. It is infinitely rich in relations, mirroring what is above, below, and what is below, above. It is animated, full of spirit and alive [...]"

Adolf Behne

As with tiles, we tend to think of glass in terms of the possibilities that it might offer. Adolf Behne's quote summarizes those possibilities of glass as a material in the best possible way: transparency and reflections. Transformation of the material that echo changes in the atmosphere, the weather or the season. An almost immaterial quality that can be found between absence and presence. Maybe one of the main properties of glass is exactly this: it can occupy – with the proper design decisions – any place on the continuum that connects some of the most obvious architectural oppositions: transparency and opacity, presence and absence, visibility and obstruction of vision; glass can acquire virtually all possible points in between those extremes and therefore generate very complex conditions.

OFIS arhitekti

One strength is the diversity of natures or physical presences it can acquire. Fabricating a glass element is a very complex process, but at the same time it is possible to create tailored elements that can suit an enormous variety of environments, conditions and atmospheres.

Something that comes to our mind as a weakness of glass is the necessity to use auxiliary materials to install it. There are always limitations on the need to use profiles, substructures, limit the sizes of the panels, create joints, etc.

OOIIO Architecture

Strengths: There is no material like glass, it is a must for most of constructions to solve interior-exterior situations. It is 100 recyclable. Provides magic games with natural light with reflections and nuances.

Weaknesses: In hot sunny weathers could be a problem to keep the interior fresh. Fragile. It is a cold material.

SLOT STUDIO

It is very common, and yet it keeps surprising us. Perhaps for its powerful visibility or our constant aim for transparency. There is something religious about glass, it is always associated with light and cleanliness. It appears to be the closest thing to nature (although it is not). Glass will always be appreciated if not needed because it closer to shelter than it is to ornament.

SMAR Architecture Studio

Glass weakness is always related to the cost. I'd say that the Strengths are related to the material properties. Of course, we have to talk here about transparency, reflection and light but glass is also able to be structural. The new Steve Jobs theatre is a recent example of how glass can become the structural element with no frames or pillars. New technologies allow glass also to have insulating properties using very thin layers of low iron glass and argon. We are now developing glass panels of 8m of height for the Science Island Museum we are designing in Kaunas. Lithuania has a very cold climate and we won't have more than 4.5 cm of glass façade.

Stefano Corbo STUDIO

Defining pros and cons in the use of glass is difficult, as every material should be considered in relation to the overall logic of the project, and based on the relationship between its different components.

stpmj Architecture

Strength: Transparency for views and the light, opacity for privacy, it can control the opacity with frit or print pattern.

Weakness: cost / price / Fragility

Studio Farris Architects

Strength: transparency and translucency, light weight, hard and durable, reusable and recyclable.

Weakness: costly material, requires regular cleaning.

SUPA architects schweitzer song

Glass doesn't have any weakness, frames have. A frame always feels like a compromise.

TAKK Architecture

The properties of transparency, flexibility, capability of changing of shape, make glass a unique material. Of course some weaknesses of it are its fragitily, price and thermal behavior.

TheeAe Architects

It goes beyond our profession to mention about technical terms of glass materials. However, for the aspect of architectural design, I don't have much ideas on it.

TOUCH Architect

Attribute of glass is mainly transparency and translucence. It is the only one building material which allows the light passing by. It is not only creating an aesthetic to the façade, but also has been used for connecting indoor and outdoor space together, as well as allows for natural sunlight to come in, which consume less of indoor lighting electricity. However, using all glass wall is not suitable for hot climate since it absorbs heat and act as greenhouse effect which increase energy consumption by using more air condition. Nor, it is easily break which is not well for security condition.

On the other hand, there is some types of glass which solves the above mentioned disadvantages, for instance, Low-E glazing is a high thermal insulated glass which allows only natural sunlight, not the heat, and, there is a safety glass which cannot be broken. However, these further features come with higher cost.

UNStudio

Philip Johnson's Glass House has already proved that a whole house can be built from glass, so strength is no longer the issue it once was. However too much glass also raises questions regarding sustainable heat loads.

Q1
본인이 유리를 사용한 프로젝트 중 가장 마음에 드는 작품 또는 다른 건축가의 작품을 소개해 달라.
(인테리어, 파사드 등 어디에 사용했든 관계없다.)

ARPHENOTYPE

현대 건축의 디자인 요소로서 유리는 거의 필수 불가결하다. 제조 및 가공 기술의 지속적인 개발과 기성 제품의 지속적인 개선이 중요한 역할을 한다. 군용 벙커와 같은 특수 건축물을 제외하고 유리는 모든 건물에 사용된다. 유리로 된 유명한 건물 중 하나는 미국 건축가 필립 존슨(Philip Johnson)이 디자인한 유리 집이다. 그는 큰 방 하나로 구성되고 완전히 유리로만 만든 집을 원했다. 이 방식으로 그는 내부와 외부 사이의 경계를 깨고 싶어 했다. 이 특성은 나를 유리(섬유)를 기반으로 한 새로운 건축 양식으로 이끈다 - 월드 와이드 웹(World Wide Web). 대부분은 이를 건물로 보지 않지만, 나는 우리의 삶의 방식을 결정하는 구조라고 생각한다. 존슨의 유리 집과 마찬가지로 내부와 외부의 경계를 깨고 결론적으로 정부에 의해 통제된다. 그러나 도시의 모든 건물을 연결하는 하수 시스템과 마찬가지로 WWW는 모든 건물을 가정 광가입자망 (Fiber to the Home, FTTH)을 통해 글로벌 네트워크의 도시로 연결한다. 그러므로 WWW는 건축이며 우리는 유리(와 데이터)에 기반한 이런 도시에서 계획하고 디자인하는 건축가로서 새로운 의식을 키울 필요가 있다.

AZC

우리는 파리에 있는 'NDBS (Notre-Dame de Bon Secours)'라는 의료시설과 프랑스 쉬렌에 있는 페이스마(Pacemar)의 사무건물 (Louis Dreyfus Armateurs Headquarters)에 유리를 썼다. NDBS 프로젝트에서 유리는 마감재이다. 페이스마의 건물에서 유리는 환기가 되는 이중 파사드 재료이다.

BOARD

우리는 최근 독일 괴를리츠(Görlitz)에 있는 지구 사무소를 확장하기 위한 건축 공모전에서 마지막 단계까지 남았었다. 우리는 이 프로젝트를 "높음과 건조함(High and Dry)"이라 불렀고 유리의 사용이 프로젝트의 성공에 결정적인 역할을 했다. 건물 전체 내에서 손님과 직원이 방향을 쉽게 찾을 수 있게 유리로 연결된 1층을 만들었다. 유리의 가장 위대한 성질과 장점을 여기서 이용해 행정 캠퍼스의 다른 모든 건물에 배리어 프리 액세스를 제공할 수 있었다. 시각적으로 투명하되 건물의 외부와 내부 사이에 충분한 분할을 제공하는 점이다. 이 투명하고 연결하는 1층을 통해 우리는 3개의 윙과 3개의 다리가 있는 블록 안의 길게 된 행정 건물을 만들 수 있었고 확장된 지구 사무소의 모든 미래 부서는 입구부터 "높고 건조하게" 접근할 수 있을 것이다.

Carlos Lampreia

유리를 투명한 벽으로 사용하는 좋은 예는 미즈 반 데어 로에(Mies van der Rohe)가 바르셀로나에 지은 독일 파빌리온(Deutsch Pavilion)이다. 이곳에서 안뜰과 같은 정사각형이나 자연이 있는 곳이면 어디에서나 외부를 향한 필터로서 유리의 능력을 느낄 수 있다. 두 경우 모두 유리는 매우 강한 연속적인 느낌을 줄 뿐만 아니라 매우 아늑한 내부 공간을 만든다.

Casanova+Hernandez Architects

알바니아 국립 사진 박물관 마루비의 경우, 우리는 건물의 강한 정체성을 정의하는 데 도움이 된 커튼월을 디자인했다. 이는 건물 중앙 안뜰의 세 정면을 덮는다.

특별히 설계된 구조 덕분에 구조가 내장되어 눈에 보이지 않고, 유리 파사드의 파티션은 리드미컬한 패턴을 따른다.

덕분에 유리 파사드는 구조적 요건이 없는 것처럼 보이며 그 패턴은 유리 파사드뿐만 아니라 건물 나머지와도 풍부한 대화를 만드는 데 표현적으로 사용할 수 있었다.

안뜰은 '음악적' 건축 요소가 되고 유리 파사드로 포착된 변화하는 자연광에 잠긴다.

유리 파사드의 투명성을 통해 밤에는 건물 내부의 조명이 안뜰의 공간을 감싸며, 휴식을 취하고 서서히 약해지는 석양의 빛을 즐길 수 있는 친밀한 장소를 만든다.

CEBRA(Mikkel Frost)

우리 프로젝트에서 유리는 창문에서 발견할 수 있다. 우리는 유리를 많이 사용하지 않는다.

25년 전 학생일 때 파리에서 메종 드 베르(Maison de Verre)를 방문했던 기억이 난다. 이는 나에게 깊은 인상을 남겼다. 이 집은 1928년 피에르 샤로(Pierre Chareau)가 설계했으며 투명 효과, 디테일 및 미학 측면에서 수십 년 앞서있었다. 오늘날에도 대중에게 공개된 것 같고 확실히 방문할 가치가 있다고 생각한다.

Davide Macullo Architects

유리는 천연 물질이다. 가장 첫째로 유리는 내부와 외부 사이의 투명성을 나타내며, 두 번째로 반사 특성을 위해 사용할 가능성이 있다. 최근 흥미로운 유리 프로젝트는 장 누벨(Jean Nouvel)이 파리 카르띠에 재단의 입구 벽에 한 작업이다. 두 가지 조건의 두 야외공간 사이의 차원과 관계가 흥미롭다. 마치 거리 정원 같다. 무자노(Mazzano)에 있는 우리 집에 좀 더 작고 친밀한 공간에서 같은 원리를 사용했고 여기서 유리는 반사적 특성을 위해 쓰였다. 둘 다 예술가 댄 그레이엄(Dan Graham)의 작품에서 영감을 얻었다.

Donner Sorcinelli Architecture

유리의 사용은 항상 투명도와 반사 사이에 생기는 흥미진진한 문제를 나타낸다. 이러한 측면에서 유리를 다른 재료와 결합해 사용했던 프로젝트 중 하나는 이탈리아 베네토 카르보네라(Carbonera)의 초등학교 디자인이었다. 이 경우 풍경과 공용 공간 및 교실 간의 시각적 연결이 전체 설계 과정의 주요 동인이었다.

Katsutoshi Sasaki+Associates

미스 반 데어 로에(Mies van der Rohe)의 독일 국립 미술관.

Keiichi Hayashi Architect

클린 룸/ 타케오카 유지

예술적인 맥락을 제외하더라도 나는 이 작품을 좋아한다. 유리 상자 안의 빈 공간, 그리고 관중 및 다른 전시작품이 유리 외관에 반사되어 보이는 비늘 무늬 공간과 같은 두 개의 공간을 동시에 만들기 때문이다.

LANDÍNEZ+REY arquitectos

유리 건축에 대한 우리의 선례는 의심할 여지 없이 건축 역사 안에서 찾을 수 있다. 우선, 1914년 피에르 샤로(Pierre Chareau)의 메종 드 베르(Maison de Verre)와 브루노 타우트(Bruno Taut)의 독일 베르크분트(Deutcher Werkbund)를 위한 파빌리온의 급진적이고 현명한 사용법은 유리를 쓰는데 필요한 건축적 교훈이다. 이 시대 이후로, 미스 반 데어 로에(Mies van der Rohe)의 건축이나 1949년에 지어진 필립 존슨(Philip Johnson)의 글래스 하우스(Glass House)에서 오늘날 세지마 카즈요의 유리 사용법까지 발전한 것 같다.

스페인 마드리드에 있는 우리 사무소의 리바스-푸투라(Rivas-Futura) 기차역은 확실히 연단에 배열된 유리 상자로, 여기 기후에 맞춰 유리 위에 떠 있는 가리개로 보호되어 있다. 유리 앞의 선형 알루미늄 루버가 친환경 그늘 덮개를 만든다. 이 루버는 또 서쪽 태양의 직사광을 막아 과열을 방지하고, 눈부심을 막으면서 내부에 빛을 반사하여 자연조명 효과도 만든다. 또한 공공 공간 사이 외부 계단이 있는 곳에 지붕이 있는 외부 영역도 만든다.

m artı d mimarlık

지질공학자 협회의 이즈미르 회의소는 우리가 좁은 구획에 설계한 소규모 프로젝트이며 부착된 구획에 있다. 서비스, 순환 요소는 구획의 뒷쪽에 있어 건물은 파사드 하나에서 채광을 받는다. 이 채광을 최대한 활용하기 위해 그 파사드에 유리 커튼월을 사용했다. 이 협회의 정체성을 반영하기 위해 파사드의 기하학에 지형학적 요소를 사용했다.

modostudio

지어지지 않은 건축 아이콘으로 미스 반 데어 로에(Mies van der Rohe)의 유리 고층건물(Glass Skyscraper, 1922)이 있다. 미스 반 데어 로에는 유리의 시학을 이해하고 이 재료를 극한으로 발전시킨 최초의 건축가 중 한 명이다.

Mork-Ulnes Architects

미스(Mies)가 생각나지 않기 어렵다. 판스워스 하우스(Farnsworth House)든 프리드리히스트라세(Friedrichstrasse) 고층빌딩 프로젝트든 근대주의에서 그는 가장 획기적으로 유리를 사용했다.

murmuro

우리는 사무실 리노베이션을 했었는데 사실 세 곳의 작은 사무실 공간을 하나의 큰 공간으로 완전히 바꾸는 일이었다. 이 프로젝트에서 우리는 필요한 흐름과 개방된 느낌을 유지하게 해 주는 유리를 칸막이로 썼다. 대략 40명 정도의 인원을 위한 메인 워크룸은 바깥을 향하는 한 면이 전부 유리와 창문틀이었고 다른 작은 두 면 역시 유리를 썼다. 작은 유리 중 한 면은 부엌과 라운지를 향했고 다른 한 면은 파트너의 사무실과 회의실을 향했다.

NISHIZAWA ARCHITECTS

벤 탄 (Ben Thanh) 레스토랑은 내부 칸막이로 유리를 사용한 가장 좋아하는 프로젝트이다.

object-e architecture

키프로스 의과대학 공모전 프로젝트는 다른 대학 건물처럼 학생들이 수업에 참석하고 바로 떠나기 위해 가는 곳이 되어서는 안 된다는 신조로 디자인했다. 대신에 학생들은 물론 강사와 행정 직원까지 하루의 중요한 부분을 소비하는 곳이므로 주요 활동뿐만 아니라 여가, 회의 및 휴식을 위한 공간에서도 고품질 공간을 제공하는 장소로 생각했다. 그런 면에서 이 디자인은 회랑의 구성을 되풀이한다. 건물의 중심에 큰 열린 공간을 만들 수 있도록 건물을 사이트의 네 가장자리까지 확장했다. 층으로 구성되고 광범위하게 식물이 심어진 이 공간이 건물의 핵심이다. 안뜰은 크기가 크고 건물의 주요 구성 요소이지만 외부에서는 보이지 않는다. 디자인의 목표는 이 안뜰이 건물에 접근하는 방문객에게 뜻밖의 발견이 되도록 비밀 공간이 남는 것이다. 반면에 건물의 외관은 '단일체'이다. 내부의 안뜰을 볼 수 있는 몇 개의 작은 구멍만 남고 그 입면에 나타나는 일부 식물이 내부에 무엇이 존재하는지에 대한 힌트를 제공한다.

교실, 실험실 및 사무실은 건물의 바깥 둘레에 배치했고 내부는 투명 유리 패널을 통해 안뜰을 마주 보고 있는 큰 복도로 점령된다. 따라서 아트리움을 마주 보고 있는 건물 측면의 내부 입면은 투명하다. 건물에서 안뜰 쪽으로 유리로 된 입면이 열리며 아트리움과 건물 내부를 일정하게 시각적으로 연결한다. 유리는 리드미컬한 방식으로 구성되지만 다양한 멀리온(mullions)으로 반복되지 않는 패턴을 만든다. 유리의 투명도는 충분한 투명도가 요구 되지 않는 곳에서 자라나는 식물을 통해서만 조절된다.

OFIS arhitekti

지금까지 우리가 유리를 사용했던 가장 흥미로운 프로젝트는 스페인의 사막에 지은 유리 집이다. 건물 외관의 수직 유리 패널을 구조로 사용하여 사막의 고속 바람을 막고 나무 외피로 된 지붕과 단을 받친다. 열 성능이 효율적인 유리 외피는 삼중 유리 벽으로 구성되어 있으며 거의 보이지 않는 코팅을 사용하여 태양으로부터 내부를 보호한다.

OOIIO Architecture

유리는 우리 프로젝트의 주요 재료가 된 적이 한 번도 없다. 물론 우리가 한 모든 건물에서 창문, 칸막이 등에 유리를 사용해왔다. 하지만 지금은 프로젝트를 정의하는 주요 재료가 아니다.

나는 이게 지금까지 우리 사무실이 있는 스페인 중부지방 같이 여름철에 특히 태양열이 굉장히 강한 곳에서 일해왔기 때문이라고 확신한다.

그런 지역에서 유리는 적절한 재료가 아니다. 또한 지역의 전통적인 건축은 유리와 같이 가볍고 맑은 것보다 불투명하고 무거운 재료에 더 기반을 두고 있다.

Piuarch

각기 다른 경사가 있는 유리, 색이 있는 유리, 태양을 반사하는 특정한 필름이 있는 유리, 매우 큰 패널 또는 얇은 시트 등 우리 프로젝트는 이 재료를 위한 실제 매뉴얼이다. 어떤 환경에서든

일정한 자연광을 위해 널리 사용했다.

상트 페테르부르크의 중심부에 있는 콰트로 코르티(Quattro Corti) 프로젝트는 확실히 언급할 가치가 있다. 기존 두 건물의 전통적인 파사드는 각기 다른 네 가지 색상이 특징인 네 개의 안뜰을 은폐한다. 빨간색, 녹색, 회색 및 파란색은 각기 다른 각도에 위치한 유리 요소로 만들어진 각 안뜰의 외관이 있고 여기서 유리는 굉장한 만화경 같은 효과를 낸다.

러시아 크라스노다르(Krasnodar)의 예카테리넨스키 의회 센터도 실험적이고 혁신적인 유리 사용법을 제공하며 현재 건설 단계에 있다. 쿠반 강을 마주보고 있는 파사드는 특별한 처리가 된 유리와 함께 물결을 반영하는 곡선 프레임으로 구성된다.

SLOT STUDIO

유리가 가장 적게 사용된 재료 임에도 불구하고 유리는 우리 프로젝트 피살(Pisal)의 주인공이다. 거대한 콘크리트 구조물이 예술적인 바닥을 내려다 보도록 특별히 설계된 유리 전망대를 감싸게 되어있다. 마치 거주 할 수 있는 만화경처럼.

SMAR Architecture Studio

SMAR의 현 프로젝트인 "리투아니아 과학 섬 및 혁신 센터"는 현재 2020년에 완료될 예정이며 투명성을 기념하는 프로젝트이다. 가로1.60m, 세로 8m의 매우 하얀 저철분 유리 패널이 있는 300m 길이의 유리 벽은 자연과 대화하는 파사드를 만든다. 반면에 지름 27m의 안뜰 세 곳은 건물의 공간을 분리하고 방문자에게 외부, 박물관 내 활동 및 전시 공간과 시각적 접촉을 항시 제공한다. 이 프로젝트에서 가장 흥미로운 점 중 하나는 내부와 외부 사이에 친밀한 관계를 형성하여 방문자에게 하늘 아래에서 걷는 느낌을 주는 동시에 섬의 녹지 분위기를 느낄 수 있게 하는 것이다.

Stefano Corbo STUDIO

메종 드 베르(Maison de Verre)는 1928년부터 1932년까지 프랑스 파리에 피에르 샤로(Pierre Chareau)가 디자인한 프로젝트이다. 유리 벽돌은 건물의 파사드에 균질하게 반투명한 벽을 만들며 투명함과 공기 순환을 위한 특정 개구부가 있다.

밤에 조명이 켜지면 집은 등대 역할을 하고 비물질화된 덧없는 물체로 보여진다.

stpmj Architecture

Glass Farm/MVRDV-brick printed glass construction

글래스를 개념화하는 작업이 가능했다

Studio Farris Architects

우리는 플랑드르 의회 파빌리온, 앤트워프에 있는 공원 타워, 그리고 브뤼헤에 있는 도시 도서관에 유리를 사용했다.

SUPA architects schweitzer song

우리 생각에 건축에서 유리는 거의 '재료'라고는 할 수 없지만, 건축에서 불가피하게 필요하다고 본다. 다시 말하지만, 우리에게 유리는 실험 파빌리온처럼 구조적으로 사용되거나 프레임이 없는 경우 가장 매혹적이다.

TAKK Architecture

SANAA가 지은 톨레도(Toledo) 미술관의 유리 파빌리온이다. 이 건물은 유리를 사용하는 것만으로 이 재료의 모든 특성을 느끼게 하는 건물이라고 생각한다.

TOUCH Architect

카페 픽셀은 태국 우돈타니의 도심에 있으며 대로에서 바로 접근할 수 있는 토지 면적이 제한되어 있다. 이 한 공간에서 카페, 레스토랑과 바가 시간별로 동시 운영된다. 아침에는 카페/빵집 겸 건강한 음식을 파는 레스토랑이며 저녁에는 수제 맥주 바가 된다.

건물의 용도가 시간별로 나뉘므로 이 카페의 용량감 및 평면은 이 두 가지 대조되는 기능에 맞게 만들어졌다. 카페 픽셀은 층이 두 개가 있는데 1층에는 카페와 바 카운터가 있고 중이층에는 고객을 위한 좌석만 있다. 이 모두 건물 법규, 규칙 및 규정에 맞는 직사각형 상자에 맞게 디자인됐으며 좌석의 양을 극대화하기 위해서 오픈 플랜과 유연한 좌석 배치가 필요했다.

단순한 상자 하나가 1x1미터의 픽셀로 채워진 픽셀 상자로 바뀐다. 이는 파사드 패턴을 만들고 내부 공간으로 이어진다. 픽셀은 파사드뿐만 아니라 픽셀 바의 좌석 및 음료 품목을 장식할 수 있는 선반으로 사용된다.

픽셀의 재료는 전부 유리이다. 각기 다른 종류와 단일 혹은 이중 유리를 통해 다양한 효과를 나타내기 위해서 투명과 반투명 유리를 모두 썼다. 흰색 반투명 유리는 낮에 카페에 들어오는 직사광선을 줄이고 반사하는데 도움이 되며, 밤에는 조명이 수제 맥주 바를 빛낸다. 카페 픽셀은 다른 시간대에 따라 다른 분위기를 만들어 낸다.

UNStudio

네덜란드 알미르(Almere)에 있는 라데팡스(La Defense) 사무용 건물의 파사드에 여러 가지 색상의 호일이 통합 된 유리 패널을 썼다. 시간 및 시각에 따라 다양한 색상이 반사된다. 건물에서 근무하는 사람들은 마치 일년 내내 매일 안뜰에서 태양이 빛나는 것 같다고 말했다.

Q2
본인이 생각하는 유리의 장단점에 대해 얘기해달라.

ARPHENOTYPE

유리의 원료는 사실상 무제한이다. 하지만 구체적인 목적으로 만들지 않으면 기계적 스트레스에 상대적으로 민감하다. 유리의 단점은 비용일 것이며 특히 특수 유리 구조일 때 더 그렇다.

AZC

장점은 투명도, 밝기와 관리 없이 오래가는 점이다.

단점은 가격과 뜨겁거나 추운 일부 극한 온도 지역에서 내부를 보호하지 못한다는 점이다.

BOARD

건축적인 면에서 끊임없는 균열의 위험에 노출돼있는 다공성 외에, 유리의 상당한 단점 중 하나는 빛과 열을 잘 전달하는 능력이다. 너무 얇거나 보호되지 않으면 맑은 날에 내부 공간을 쉽게 과열할 수 있다. 그런 일이 일어나지 않도록 우리는 "높음과 건조함" 프로젝트의 유리 파사드 전체에 그늘이지는 차양으로서 알루미늄 틀을 설치할 것을 제안했다. 삼중 유리 역시 여름철 내부 공간에 적당한 온도를 유지하는 데 도움이 되었다. 또한 스테인리스 사지 직조로 만든 매우 효과적인 가림막은 더운 날에도 쾌적한 실내 온도를 보장한다. 이러한 모든 방안으로 에너지 효율이 높은 행정 건물을 만들었다.

Carlos Lampreia

유리는 놀라운 물질이다. 왜냐하면 유리를 사용하여 완전하게 투명한 벽을 지을 수 있기 때문이다. 하지만 가끔 유리를 보이지 않는 물질로 생각하는 오류를 저지르는 데 그것은 실수다. 유리는 완벽하고 빛나는 매우 가공된 물질이기 때문에 모든 재료 중에서 가장 눈에 띈다. 더 큰 단점은 그 취약성과 유리를 사용하는 데 필요한 하부 구조에 대한 의존성이다.

Casanova+Hernandez Architects

앞서 언급했듯이, 유리의 잠재력은 엄청나다.

유리는 투명하거나 다양한 반투명성을 나타낼 수 있다. 색상을 넣을 수 있고, 패턴을 인쇄할 수 있으며, 반사가 풍부하거나 아예 무광일 수도 있다.

요즘 디자이너는 유리의 여러 가지 특성을 사용하거나, 다른 성질의 유리를 결합하거나, 곡선을 만들거나, 다른 재료와 대조를 만들어 매우 다른 건축적 의도를 표현할 수 있다. 우리 생각에 유리는 많은 재료 중 가장 다재다능한 재료이다. 유리는 독특한 방식으로 시적 의도를 건축으로 번역한다.

CEBRA(Mikkel Frost)

덴마크의 건축법은 에너지 효율에 관해서 다소 엄격하다. 그래서 우리는 큰 유리창을 설계할 때 겨울에는 열손실로, 여름에는 과열로 고생한다. 하지만 클래딩 재료로서 유리는 상당히 훌륭하다. 우리는 그룬드포스 전문대(Grundfos Kollegiet)의 거대한 거울 공간에 유리를 사용했다. 재밌는 점은 건축가가 완전히 피할 수 없는 유일한 재료는 유리일 것이라는 점이다. 아마 소금 없이 요리하는 것과 같을 것이다.

Davide Macullo Architects

유리는 경이로운 재료이다. 가장 견고하거나 가장 존재하지 않는 재료, 혹은 가장 도전적이고 가장 위험한 재료가 될 수 있다. 너무 많은 반사와 투명도는 인간의 본성에 반한다. 건축 산업에서 유리를 광범위하게 사용하면서 사람들이 받는 스트레스를 높이는 일련의 불편한 상황이 생겨났다. 나는 대도시 지역의 유리 건물과 같은 상황을 말하는 것이다. 예를 들어 런던의 사무 공간은 인간의 근무 환경에 너무 많은 유리의 사용으로 고통받고 있다.

Donner Sorcinelli Architecture

유리에 장점이나 단점은 없다. 오직 시적 잠재력뿐이다.

Katsutoshi Sasaki+Associates

단점은 단열 성능이다. 장점은 많이 있다.

Keiichi Hayashi Architect

장점: 투명한 재료이다.

단점: 너무 강하고 자각없이 내부와 외부가 너무 갑자기 합쳐진다.

LANDÍNEZ+REY arquitectos

건축에서 유리 사용의 장단점은 투명성을 둘러싼 본질적인 가치 또는 빛에 반응하는 특성과 관련이 있을지도, 없을지도 모르겠다.

오늘날 건축에서 유리 사용에 대한 고전적인 담론은 에너지 요인이 필수적으로 추가되었다.

m artı d mimarlık

유리는 투명한 특성이 장점이지만 부족했던 단열 및 방음 특성은 기술적 개발로 향상했다. 유리를 간격을 두고 층층으로 사용하고 다양한 재료로 코팅하면 단열성이 향상된다. 결과적으로 유리를 사용하기 더 쉬워지며 더 널리 보급된다.

modostudio

유리의 장점은 투명성이지만, 이는 단점이 될 수도 있다. 유리는 극도로 주의를 기울여 사용해야 하는 재료이며 신중하게 기술적으로 디테일을 디자인해야만 우아함을 지닐 수 있다.

Mork-Ulnes Architects

유리는 더 개방적이고 가벼운 구조물을 지을 수 있는 더 많은 기회를 열어준다.

우리에게 유리의 단점은 유리를 너무 많이 사용하는 경우이다. 너무 많은 투명성을 만들고 싶은 유혹이 흔하며, 이는 전망을 그리고 빛을 조절할 힘을 앗아간다.

murmuro

유리는 정신적 장벽을 정의하는 능력이 있지만 동시에 정의하지 않기도 한다. 유리의 단점은 청각적 편안함을 이루기 어렵다는 점일지도 모른다.

NISHIZAWA ARCHITECTS

장점

중간 공간을 만들고 내부와 외부 공간을 연결하는 데 도움이 된다.

투명하고 반사되는 특성은 햇빛이 유리에 반사 될 때 그 각도로 각기 다른 시간, 날씨 또는 계절에 따라 주변 상황에 맞는 흥미롭고 다양한 광학 효과를 만들 수 있다. 칸막이가 모든 사람들에게 건물에 올 때마다 새로운 느낌을 줄 수 있다면 좋겠다고 생각한다.

단점

때로는 가장자리와 표면을 보호하기 위해 프레임이 필요하다. 이 경우, 유리는 수공예품 같아서 건물에 현실로 만드는데 시간과 돈이 많이 소요된다.

내부의 시원한 공기가 빠져 나가는 것을 완전히 막을 수 없다.

쉽게 더러워지기 때문에 정기적으로 청소해야 한다.

object-e architecture

"순도와 선명도, 빛나는 가벼움과 투명한 정확성에 대한 열망, 물질적이지 않은 가벼움과 무한한 활기에 대한 갈망은 유리에서 이를 성취할 수단을 발견했다. 가장 형언할 수 없고, 가장 기본적이고, 가장 유연하고, 가장 변화할 수 있는 재료로, 의미와 영감이 가장 풍부하며 그 어떤 재료와도 다르게도 주변 세계와 융합된다. 이 가장 불변의 재료는 대기의 모든 변화에 따라 변한다. 위에 있는 것을 아래로, 아래에 있는 것을 위로 반사하여 관계가 무한히 풍부하다. 유리는 생기 있고, 활기차며 살아있다 [...]"

아돌프 베네

타일과 마찬가지로 우리는 유리가 제공할 가능성의 관점에서 유리를 생각하는 경향이 있다. 아돌프 베네의 인용문은 유리의 가능성을 가능한 가장 좋은 방법, 즉 투명성과 반영으로 요약한다. 재료가 대기, 날씨 또는 계절의 변화를 반영하여 변형하고 부재와 존재 사이에서 발견할 수 있는 거의 비물질적인 특성을 지닌 것이다. 유리의 주요 특성 중 하나는 정확히 다음과 같다. 투명도와 불투명도, 존재 및 부재, 시야 및 시야 방해와 같은 가장 명백한 반대 요소를 연결하는 연속체의 어느 곳이든 차지할 수 있다는 점이다. 유리는 이러한 극단 사이의 거의 모든 가능한 점을 획득할 수 있으므로 매우 복잡한 조건을 만든다.

OFIS arhitekti

한 가지 장점은 유리로 얻을 수 있는 특성이나 물리적 존재감이 다양하다는 점이다. 유리 요소를 제작하는 것은 매우 복잡한 과정이지만 동시에 엄청나게 다양한 환경, 조건 및 대기에 적합한 맞춤형 요소를 만들 수 있다.

유리의 단점으로서 생각나는 것은 설치를 위해서 유리 외에 추가로 재료를 사용해야 한다는 점이다. 종단면, 하부 구조, 패널 크기 제한, 이음매 등 때문에 항상 제한이 있다.

OOIIO Architecture

장점: 유리와 같은 재료는 없으며 내부-외부 상황을 해결하는데 대부분 건물에 필수불가결하다. 100% 재활용이 가능하다. 반사와 뉘앙스가 있는 자연광으로 마법 같은 효과를 부린다.

단점: 더운 날씨에는 실내를 쾌적하게 유지하는 것이 어려울 수 있다. 부서지기 쉽고 차가운 물질이다.

SLOT STUDIO

유리는 매우 흔한 재료지만, 우리를 계속 놀라게 한다. 아마도 유리의 강력한 가시성이나 투명성을 향한 우리의 지속적인 목표 때문일지도 모른다. 유리에는 종교적인 무언가가 있다. 항상 빛과 청결함을 연상하게 한다. 자연에 가장 가까운 그 무언가로 보인다 (실제로 그렇지는 않지만). 유리는 장식보다 거처에 더 가깝기 때문에 필요하지 않더라도 항상 환영받을 것이다.

SMAR Architecture Studio

유리 단점은 항상 비용과 연관된다. 장점은 물질적 특성과 관련이 있다고 말하고 싶다. 물론 유리의 장점을 논할 때 투명성, 반사, 빛에 관해 이야기해야 하지만 유리는 구조가 될 수도 있다. 새로운 스티브 잡스 극장은 유리가 어떻게 프레임이나 기둥 없이 구조가 될 수 있는지 보여주는 최근 사례이다. 새로운 기술 덕분에 아주 얇은 저철분 유리와 아르곤 층을 사용하여 유리에 단열성이 생겼다. 우리는 현재 카우나스에서 디자인하고 있는 과학 섬 박물관을 위해 높이 8m의 유리 패널을 개발 중이다. 리투아니아는 기후가 매우 추우며 이 프로젝트에서 두께 4.5cm 이상의 유리 파사드는 쓰지 않을 것이다.

Stefano Corbo STUDIO

유리에 대한 장단점을 정의하는 것은 어렵다. 모든 재료는 프로젝트의 전반적인 논리는 물론 다른 요소 간의 관계에 따라 고려되어야 하기 때문이다.

stpmj Architecture

장점: 채광이나 조망을 위한 투명함, 반대로 프라이버시를 위한 불투명함. 단일 재료 임에도 투명함 정도를 여러 테크놀로지를 통해 조절이 가능하다. Frit이나 프린트 패턴의 다양함 확보

단점: 예산에서의 비중이 크다 (코스트). 깨지기 쉬워 유지보수에 비용이 발생한다.

Studio Farris Architects

장점: 투명성과 반투명성, 경량, 단단함과 내구성, 재사용 및 재활용 할 수 있다는 점이다.

단점: 가격이 비싸며 정기적인 청소가 필요한 점이다.

SUPA architects schweitzer song

유리에 단점은 전혀 없다. 창틀은 언제나 타협처럼 느껴진다.

TAKK Architecture

투명성, 유연성과 모양을 변경할 수 있는 특성 덕분에 유리는 독특한 재료이다. 물론 단점은 부서지기 쉽다는 점과 가격 및 열적 거동이다.

TheeAe Architects

유리 재료의 기술적 특성에 대해 언급하는 것은 우리 분야 바깥의 일이다. 하지만 건축 디자인의 측면에서 봐도 별 다른 생각이 없다.

TOUCH Architect

유리의 주된 속성은 투명성과 반투명성이며 빛이 지나가는 것을 허용하는 유일한 건축 자재이다. 파사드에 미학을 만들어 낼뿐만 아니라 실내와 실외 공간을 연결하는 데 사용되며 자연채광 덕분에 실내조명에 전기를 덜 소비할 수 있다. 그러나 유리는 열을 흡수하고 온실 효과를 일으킨다. 이 때문에 더 많은 에어컨을 사용하게 되고 에너지 소비를 증가시키기 때문에 모든 벽에 유리를 사용하는 것은 더운 기후에 적합하지 않다. 또한 쉽게 깨지기 때문에 보안을 위해서도 그다지 좋은 재료는 아니다.

반면에 위에 언급한 단점을 해결하는 유리 종류도 있다. 예를 들어, 열을 제외한 자연광만을 허용하는 저방사 유리로 로이유리(Low-E)가 있으며 깨지지 않는 안전유리도 있다. 그러나 이러한 부가 기능은 비용이 많이 든다.

UNStudio

필립 존슨 (Phillip Johnson)의 글래스 하우스(Glass House)가 집 전체를 유리로 지을 수 있다는 것을 이미 증명했기 때문에 내구력은 더 이상 예전처럼 문제가 아니다. 그러나 너무 많은 유리는 지속 가능한 열 부하에 관한 의문을 제기한다.

Interviewee
PROFILE

ARPHENOTYPE

Dietmar Köring, Dipl.-Ing.(FH) M.Arch. Architect BDA, is an architect, researcher, and educator living in Cologne. He is head of the architectural research office Arphenotype, where he focuses on blurring the boundaries of different artistic disciplines. From 2012 to 2017 he was a research fellow at TU Berlin / CHORA City & Energy and Dietmar has taught Digital Design at TU Braunschweig from 2010 to 2012, he was Guest Professor for Virtual Realities & Experimental Architecture at the University Innsbruck ./Studio3 in 2011, Technology and Design Lecturer at the Cologne Institute for Architectural Design / C-I-A-D and visiting lecturer for digital design at the DeMontfort University Leicester. From 2011 to 2012 he was assistant professor for Smart Grid research (Smart City Concepts 2022) at the Institute for Corporate Architecture at the Cologne Technical University.

He studied architecture at the University of Applied Sciences Cologne, the University of Western Sydney and at the Muthesius Academy of Fine Arts, where he graduated as in 2005 as Dipl.-Ing. (FH). Dietmar received his MArch in 2007 at the Bartlett School of Architecture University College London, under Prof. Neil Spiller and Phil Watson. Since 2008 he is a registered Architect at the AKNW and ARB.

Through his career he has worked internationally for offices such as Coop Himmelblau, Graft, 3deluxe and Andrew Wright Associates. His research has been awarded by the Jaap Bakema Fellowship / NAI and his works have been internationally published and exhibited, including MoMa New York, Heide Museum of Contemporary Arts Australia and Deutsches Technikmuseum Berlin. Dietmar has given international lectures, guest critiques and workshops. Since 2013 he is collaborating with Simon Takaski as Takasaki Koering Architects.

Dietmar is member of the narrative research network .horizon.com.

AZC

AZC was founded in 2001 with the idea that exploring architecture and its techniques could help to improve our built environments. Our interest does not lie in inventing concepts, we have always sought to realize buildings for real life's needs.

Through competitions and direct commissions, our office has worked on over a hundred projects of varied scales and uses. Most of our built projects are intended for a wide audience; sports facilities, lecture halls, office buildings and residential, some of which very specific for vulnerable populations. We also have, eight metro stations under construction, including four in Paris and four in Rennes and studies for a new station in Lyon, are ongoing.

Through some recently completed buildings, which have different purposes, we want to share our current concerns of coherence with global and local contexts which today represent the major issues of architecture.

We are not alone in the projects process, our clients and our partners share this common experience, which is engaging and meaningful; they allow us to reflect on our own actions that relate to the projects. We aspire to a high quality in any form of collaboration.

Most of our work has been published, displayed, sometimes awarded and we have often been given the opportunity to speak on topics of sustainability, diversity and innovative techniques, which all illustrate our commitments.

BOARD

BOARD (Bureau of Architecture, Research, and Design) was founded in Rotterdam in 2005 and is active in many fi elds: as an architecture, urban design, and design practice, as a research board and as a platform for comparative analysis on urban issues through its bi-annual journal MONU – Magazine on Urbanism. BOARD won several prizes recently in prestigious international architecture and urban design competitions.

Bernd Upmeyer is the founder of BOARD and editor in chief of MONU – Magazine on Urbanism. He studied architecture and urban design at the University of Kassel(Germany) and the Technical University of Delft (Netherlands). From 2004 until 2007 he taught and did research as Assistant Professor at the department of Architecture, Urban Planning and Landscape Planning at the University of Kassel. In 2010 he taught as Adjunct Professor at the department of Urban Design at the Hafen-

City University Hamburg. In 2012 he was a guest critic at the Berlage Institute's fi rstyear postgraduate research studio "Anarcity".

In 2013 he lectured and participated in a discussion about architecture, urbanism and media at Strelka's Urban Studies Session in Moscow. Upmeyer frequently writes for international publications and magazines. He holds a PhD (Dr.-Ing.) in Urban Studies from the University of Kassel(Germany). Upmeyer is the author of the book Binational Urbanism – On the Road to Paradise. The book examines the way of life of people who start a second life in a second city in a second nation-state, without saying goodbye to their fi rst city.

Upmeyer coined the term "binational urbanism".

BOARD employs an international team of architects and planners and collaborates with national and international external consultants and specialists.

Carlos Lampreia

Carlos Lampreia, architect (1990), is architecture design teacher at FAA-Universidade Lusíada de Lisboa since 1994, studied at OPorto Architecture School and at Lisbon Technical University FA-UTL. Master in architecture theory, 'towards an objective architecture', 2002. Phd about, strategy, site and material, concerning architecture and arts, 'concept site and material, a strategy in architecture and arts, 1960-2000', 2017. His Lisbon based office, carloslampreia[x]arquitectos, works on an experimental way with young architects and students towards architectural materialisation, participating both in international competitions and individual private requests.

Casanova +Hernandez Architects

Casanova+Hernandez, founded in 2001 by Helena Casanova and Jesus Hernandez, is a design and research studio based in Rotterdam. It focuses on rethinking and designing our urban habitat in order to create vibrant cities while promoting environmental and social sustainability.

Working with an interdisciplinary team and with experience developing projects in very different cultural contexts in Europe, South America and Asia, the office has expanded its capabilities and its international network through close and fruitful collaboration with experts in different continents.

Casanova+Hernandez is structured in two complementary platforms: C+H Projects and C+H Think Tank. C+H Projects is the design platform of Casanova+Hernandez. It operates in the fields of architecture, landscape architecture and urban design, often combining them to create hybrid architectural landscapes.

C+H Think Tank works as an independent platform that analyses urban and social problems and proposes innovative design solutions, new urban strategies and advice on the implementation of new policies.

www.casanova-hernandez.com

CEBRA

CEBRA is a Danish architectural office founded in 2001 by the architects Mikkel Frost, Carsten Primdahl and Kolja Nielsen. In April 2017, architect MAA Mikkel Hallundbæk Schlesinger entered the group of partners.

Based in Aárhus in Denmark and in Abu Dhabi in the UAE, CEBRA employs a multidisciplinary international staff of 50 architects, constructing architects, urban planners and landscape architects, who all share a strong passion for architecture.

CEBRA has gained recognition through award-winning projects such as The Iceberg at the habour front in Aarhus and the Experimentarium science centre in Copenhagen and has a growing international portfolio in Europa and the MENA region.

At CEBRA we want to change the way to think, design and build architecture. We are always pushing artistic and architectural boundaries - pushing these boundaries with a CEBRA attitude and a Nordic mindset that combines our artistic approach to architecture with an understanding of its cultural context.

We design architecture by listening to and understanding our users and clients and studying their context, culture and climate. Our services cover all project phases - from client advisory and user involvement and concept and project development to project and construction management as well as technical supervision.

Most CEBRA projects are within the fields of education, culture and housing - thought, designed, and built in line with our mantra - Architecture with attitude.

Davide Macullo Architects

Davide Macullo (b. Giornico, CH, 1965) lives and works in Lugano, Switzerland. Studied art, architecture and interior design. For 20 years (1990-2010) he was project architect in the atelier of Mario Botta with responsibility for over 200 international projects worldwide. He opened his own atelier in 2000.

The ethos of the studio is one of 'drawing from context' and the various contributions promote a dialogue between the specificity of the project and the universality of the contexts. His work has been published and awarded both at home and abroad. Selected realized projects include the WAP ART foundation mixed use gallery and apartment in Gangnam Seoul, South Korea, the Assuta Hospital in Ashdod, Israel, 5* Hotel and SPA facilities in Greece,the headquarter Jansen AG in Oberriet, Switzerland, Private Museum in Jeju South Korea, Sino-Swiss centre in Tianjing China, several houses and housing in Switzerland and abroad.

Current projects include a new Health and Wellness Hotel in Weggis, Switzerland and Marbella, Spain, houses and residential buildings in Switzerland, a beachfront villa in Heraklion, Greece, a Medical SPAin Baku, Azerbaijan. The work of the studio includes masterplanning, graphic design, branding consulting and custom designed furniture, now in production and spans to the creation of contemporary art collections for clients.

In Rossa Calanca Valley in the Grison Canton, Davide Macullo has started an urbanistic program to promote the intervention in situ of international artists to influence daily life through contemporary art. The first building realized in collaboration with Daniel Buren will be followed by other ten artists.

Donner Sorcinelli Architecture

Donner Sorcinelli Architecture is an international architectural design office based in Italy.

Founded by architects Luca Donner and Francesca Sorcinelli, the firm pays particular attention to the theme of sustainable and affordable architecture in all its variants, based on experimentation and research in various fields like Architecture, Urban Design, Interior and Product Design.

Their projects have been awarded in International competitions:

"Social Housing Dev."- Piazzola sul Brenta 1st prize; " Social Housing Dev." -Presina, 1st prize; "Design Beyond East and West"- Seoul, 1st prize; "International Design Competition for Modern Saudi Houses, Affordability and Sustainability"- Riyadh, 1st prize; Urban Retrofitting of S.Elena's - Silea, 1st prize; Sansovino Masterplan -Montebelluna, 3rd prize; School Campus in Carbonera, 3rd prize;"Your Absolute" for a residential Tower, Mississauga, Honorary Mention; "Daejeon Urban Renaissance"- Daejeon, Honorable Mention.

They have been published in many international magazines, books and presented in several exhibitions in Italy and abroad.

DoSo are winners of the "Cityscape Architectural Review Award 2006", "SAIE selection Awards 2009" and the "20+ 10+ X World Architecture Award 2012". They have received an Honorary Mention at Modern Atlanta Prize 2011, an Acknowledgement Prize at Holcim Awards 2005 for sustainable constructions (MENA region) and they have been nominated by Korean Institute of Architects among "100 Architects of year 2017".

Luca Donner and Francesca Sorcinelli have been teaching at International Universities in Dubai after previous academic experiences in Italian Universities.

www.doso.it

Katsutoshi Sasaki +Associates

Office Information
4-61-3 Tanaka-cho
Toyota-shi Aichi
471-0845 Japan
Phone:+81 565 29 1521
sasaki@sasaki-as.com

1976 Born in Toyota-shi Aichi,Japan
1999 Guraduated Kindai University
2008 Established Katsutoshi Sasaki + Associates

Keiichi Hayashi Architect

Office information
1-6-10 Kumochi-cho Chuou-ku
Kobe-city Hyogo
651-0056 JAPAN
Phone +81.78.221.1868
hayashi @8107.net
Owner: Keiichi Hayashi
Established: 1997

Biography
1967 Born in Osaka
1991 Graduated from Metal Engineering, Kansai University
1993 Graduated from Architecture, Kansai University
1997 Established Keiichi Hayashi Architect

Design Philosophy
It is important for me to make architecture using basic materials and uncomplicated construction methods. I try to create a system that is based on pure architecture but becomes complex when people use it.

www.haya-at.com

LANDINEZ+REY arquitectos

LANDINEZ + REY is an architectural practice co-founded in 2000 and based in Madrid(Spain) by the architects David Landinez González-Valcácel (Madrid, 1973) and Mónica González Rey (Paris, 1973). Both are formed as M.Arch (1999) in ETSAM-UPM(Faculty of Architecture of the Polytechnical University of Madrid, UPM) and both are also graduated as Building Engineers by UEM-Madrid (2013). David also is M.Arch in Efficient Buildings and Rehabilitation by UEM (Universidad Europa de Madrid) and Mónica has also postgraduated studies in Analysis and Real Estate Management by Colmillas University (ICAI-ICADE)

LANDINEZ + REY arquitectos [el2gaa] develops its activity as a working plaform where architecture is sought from its capacity as a system generator. Systems capable of providing answer to both the place and the rest of the dimensions, scales and techniques demanded by each draft. Its work is based on the research that arises from the proposals presented to architectural competitions, many of them has been rewarded with awards and honorable mentions, others won and built.

Among these works we can find the Badajoz, Coria and Plasencia secondary schools and the Malpartida de Plasencia and Jaraíz de la Vera Gym Plvillions for the Regional Government of Extremdura, Spain and the RivasFutura Underground Station for Metro de Madrid.

His works have been published in different specialized magazines and exhibitied by different national and foreign insititutions.

www.landinez-rey.com

M artı D Mimarlık

M artı D Mimarlık was founded in 1987 in İzmir by Metin Kılıç and Dürrin Süer. They design various types of projects in various scales such as residential, commercial, healtcare, educational and urban design. With their intention that unites academic and practical skills, they contribute to today's architecture culture.

Metin Kılıç - Partner - Founder(Architect)
He was born in 1962. In 1985 he graduated from 9 September University, Faculty of Architecture. He has been working and conducting M artı D Mimarlık since 1987 as founder and partner. He has many projects sucha as hospitals, educational instutions, hospitals, residences, commercial buildings. He has been principal of İSMD between 2013-2015. He has many prizes in architectural competitions.

Dürrin Süer - Partner - Founder (Phd. Architect)
She was born in 1965 in Ankara. In 1987 she graduated from 9 September University, Faculty of Architecture. She has completed Masters and PhD Degrees in same university. Between 1987 - 2007 she has worked as academician in 9 September University, Faculty of Architecture. She has many articles published in architectural magazines about architecture education, architecture and utopia, architecture and technology, consumption spaces and residental spaces. She has been working in M artı D Mimarlık as founder and partner. She also takes place as jury member in architectural competitions and writes academical researchs and articles.

Ali Can Helvacıoğlu (Architect)
He was born in İzmir in 1989. He was graduated from Izmir University of Economics, Faculty of Architecture in 2011. He has been working in M artı D Mimarlık since 2011.

modostudio

modostudio | cibinel laurenti martocchia architetti associati, located in Rome, is a multidisciplinary practice of architecture, urban planning and industrial design.

Profiting from the diversified skills of the founding partners and the continual collaboration with experts from various fields, modostudio combines architectural theory, research, innovation and experimentation with high technical knowledge and professionalism. Established at the end of 2006 by three principal architects, Fabio Cibinel, Roberto Laurenti, and Giorgio Martocchia, after many years of collaborating with internationally acclaimed architects like Massimiliano Fuksas, Piero Sartogo, Erik Van Egeraat and Kas Oosterhuis, modostudio in a short time was awarded and shortlisted in many international architectural competitions. The office completed a 13.500sqm office building in Nola (IT) for Giorgia & Johns spa fashion company, the new headquarters for the Elisabeth and Helmut Uhl research foundation in Laives (IT) and a research building for Intecs spa company in Rome. Actually is ongoing the desing for new theater and urban park of Piazza d'Armi site in L'Aquila (IT).

Mork-Ulnes Architects

About Casper Mork-Ulnes

Norwegian born, Casper Mork-Ulnes was raised in Italy, Scotland and the United States, which has brought a broad perspective to his eponymous firm's work. In 2015, Casper was named one of "California's finest emerging talent" by the American Institute of Architects California Council. He was selected by the Norwegian National Museum as one of "the most noteworthy young architects in Norway" with the exhibit "Under 40. Young Norwegian Architecture 2013." Casper holds a Master of Architecture from Columbia University and a Bachelor of Architecture from California College of the Arts.

With offices in San Francisco and Oslo, Mork-Ulnes Architects approaches projects with both Scandinavian practicality and Northern California's 'can-do' spirit of innovation.

Rigorous and concept-driven, the practice is based on built work characterized by both playfulness and restraint, and informed by economies of means and materials. Mork-Ulnes Architects have worked on projects ranging in scale from masterplans to 100 square foot cabins, and have realized buildings on 3 continents.

Mork-Ulnes Architects has been the recipient of numerous national and international honors, including Architectural Record's 2015 worldwide Design Vanguard award.

The work of Mork-Ulnes Architects has also been widely featured in international publications such as The New York Times, Wallpaper, Mark, and Dwell.

www.morkulnes.com

murmuro

João Caldas (Braga, 1981) is the co-founder, with Rita Breda (Estarreja, 1981) of the office murmuro, working from Braga and Porto, in Portugal. They have started their joint path while still at the university, having graduated from DARQ-FCT, University of Coimbra. As Erasmus students, at the NTNU and the Fine Arts Academy in Trondheim, Norway, they had the opportunity to taken part of an interdisciplinary and collaborative program for architecture and art students, pivotal in their education.

The success of some of their projects led to the foundation, in 2015, of the office murmuro, where they develop projects regardless of their scale, program or budget. From their body of work they highlight the multifamily housing project Doze Casas, in Braga, shortlisted for the PREMIS FAD award in 2016, the VII ENOR Award in 2017, and the 2nd Prize on the Serralves Foundation's Pavilion Competition. Recently murmuro has received EUROPE 40 UNDER 40 award that spotlights the most promising and the best emerging young architects in europe.

NISHIZAWA ARCHITECTS

SHUNRI NISHIZAWA

was born in 1980. He obtained his B.Arch (2003) and M.Arch (2005) degrees from Tokyo University. He worked from 2005 to 2009 for Tadao Ando Architect & Associates (Osaka). Then, he worked with Vo Trong Nghia (Ho Chi Minh City, 2009-11). He was a Partner in Sanuki+Nishizawa Architects (2011-15), before founding Nishizawa Architects in 2015. His work includes the Binh Thanh House (with Vo Trong Nghia Architects, Ho Chi Minh City, 2013); Thong House (Ho Chi Minh City, 2014); Katzden Factory (Binh Duong City, 2016); House in Chau Doc (2017; published here); Pizza 4P's Ben Thanh (Ho Chi Minh City,2017), all the completed buildings are in Vietnam.

object-e architecture

Object-e architecture is an architectural practice currently based in Thessaloniki, Greece and directed by Katerina Tryfonidou and Dimitris Gourdoukis. It started in 2006, in St. Louis, USA, as a platform with the intention to explore new territories in architecture with the aid of computational tools and techniques. Through time, object-e architecture moved beyond the borders of computation and engaged design at large, trying to graft the new media with the social, political and ecological issues that architecture is facing today. It has won a number of prizes in international competitions, and its works has been published, exhibited and presented internationally.

Object-e architecture is based on several collaborations with people coming from different backgrounds, with different design intentions and agendas. The outcome of object-e architecture, being in most cases collaborative, is therefore defying any concept of style; Identity is formed through difference and constant transformation.

Object-e architecture engages architecture and design in three different - but always connected - levels: Through specific design projects that are answers to specific design questions; private projects or competitions. Through experimental research projects that are aiming to extend our understanding of space, digital media and fabrication. Through teaching; either in established institutions or independently, sharing of knowledge is always the core that allows the rest to happen.

http://object-e architecture.net
https://www.facebook.com/object.e

OFIS arhitekti

Based in Ljubljana, formed by Rok Oman and Spela Videcnik (1998)

Rok Oman
(born 1970) studied architecture at Ljubljana School of Architecture (grad.1998) and at Architectural Association in London(grad.2000).
Currently teaching in Harvard GSD, Boston, MA

Spela Videcnik
(born 1971) studied architecture at Ljubljana School of Architecture (grad.1997) and at Architectural Association in London (grad.2000).
Currently teaching in Harvard GSD, Boston, MA

OOIIO Architecture

OOIIO is an international team of architects, designers and engineers engaged in finding this special "I don´t know what it is" that makes a work unique, exciting and able for transmit sensations on a way that a vulgar work will never get.

In OOIIO we know that not every construction is architecture.
We do architecture.

OOIIO creative process is completely open and random. When we start a project we never know how it will end, what will come out, that's why we introduced throughout the design process countless stimulus and references from any kind that may improve the final design. These are usually unexpected and we find them everywhere, like in the geometry of a mineral that seems attractive, a traditional dish of the place where the project is built, the shape and colors of a vase, a tree or a graffiti that we found painted on a wall.

This constant search for poetical links to architecture through everything around us makes us look at the world with wide open eyes and helps us to get every project as the result of a unique creative process, making each OOIIO project special and standed out.

www.ooiio.com

Piuarch

Francesco Fresa, Germán Fuenmayor, Gino Garbellini and Monica Tricario formed the Piuarch studio in 1996 out of a desire to merge different experiences into a shared architectural project.

The studio is located in an open space in a former industrial building that once hosted a typography business in Brera, in the centre of Milan. Here, Piuarch designs public buildings, office and residential complexes, commercial spaces, boutiques, shopping malls and even urban plans, with the contribution of consultants from various disciplines.

Piuarch has pursued these themes participating in competitions, developing projects from the planning to the final construction phase, elaborating interior design projects.

In recent years Piuarch has developed a number of projects abroad. It is active in China, Algeria, Russia, where it has recently opened an operational office, and in Ukraine, with ongoing and realized projects.

SLOT STUDIO

SLOT is an active and interdisciplinary architectural design studio. People from diverse professional disciplines contribute to this project.

Our work has reached a profound understanding of the human needs, enabling us to intertwine the constructive and philosophical sides of building.

As architects, we push design to its ultimate material consequences and aim for cultural connectivity: sense of usability, mathematics of space and a wide aesthetic research.

Juan Carlos Vidals, Founder and Director of SLOT, graduated from the UNAM - National Autonomous University of Mexico in 2002 and has worked in the LCM/Fernando Romero and collaborated with Rojkind Arquitectos in several projects.

slot.mx

SMAR Architecture Studio

SMAR Architecture Studio, founded by UWA Professor Fernando Jerez (PhD MArch ETSAM) in 2009 and co-directed by Belen Perez de Juan is an awarded Western Australia and Madrid based group of architects and urban thinkers operating with architecture, technology and society. Our projects deploy near-future scenarios as critical instruments for instigating debate about urban and social issues through design and emerging technologies, in order to improve social and political interaction in relation to the environment.

SMAR, with more than 20 international awards, after winning the Science Island International Design Competition is currently building the new Science Museum and Innovation Center in Kaunas, Lithuania to be completed in 2020. SMAR has been recently shortlisted in the Guggenheim Helsinki International Competition among 1715 entries and awarded 3rd Prize in the Aalto Museum Competition, MALI Lima Museum of Art Competition among 689 entries in 2016. Their work and writings have been published in international Journals such as AV (Arquitectura Viva), Architectural Review, A'A', L'Architecture d'Aujourd'hui, Bauwelt, WA wettbewerbe aktuell, EGA, AW, Archdaily, dezeen, Designboom or The Architect (Official journal of the Australian Institute of Architects) and exhibited in the (AIA) American Institute of Architecture, New York, The 2014 National Architecture Conference of Australia or the Solomon R. Guggenheim Foundation, Kunsthalle, Helsinki.

Stefano Corbo Studio

Stefano Corbo (1981) is an Italian architect, researcher, and Assistant Professor at RISD (Rhode Island School of Design).

He holds a Ph.D and an M.Arch. II in Advanced Architectural Design from UPM-ETSAM Madrid (Escuela Técnica Superior de Arquitectura).

Stefano has taught at several academic Institutions: Nanjing University; LAU Beirut (Lebanese American University); The Faculty of Architecture in Alghero, Italy; ETSAM Madrid; he has also been a guest lecturer at SAC Städelschule Frankfurt, Deakin University in Melbourne, College of Design Minnesota, ESALA Edinburgh, The University of Miami, and The University of Wisconsin.

Stefano has contributed to several international journals and has published two books: "From Formalism to Weak Form. The Architecture and Philosophy of Peter Eisenman." (Ashgate-Routledge, 2014), and "Interior Landscapes. A visual atlas." (Images, 2016).

In 2012, after working at Mecanoo Architecten, Stefano founded his own office SCSTUDIO (www.scstudio.eu), a multidisciplinary network practicing architecture and design, preoccupied with intellectual, economic and cultural contexts.

www.scstudio.eu

stpmj

stpmj is an award winning design practice based in New York and Seoul. The office is founded by Seung Teak Lee and Mi Jung Lim with the agenda, "Provocative Realism". It is a series of synergetic explorations that occur on the boundary between the ideal and the real. It is based on simplicity of form and detail, clarity of structure, excellence in environmental function, use of new materials, and rational management of budget. To these we add ideas generated from curiosity in everyday life as we pursue a methodology for dramatically exploiting the limitations of reality.

We design identity, brand and value.
We design creative and innovative culture.
We design unique vision of architecture.

We have been recognized with architectural awards including,

Architectural Record Design Vanguard
American Institute of Architects New York Design Award
Korean Ministry of Cultre, Sports and Tourism Young Architects Award
Kim Swoo Geun Preview Award
American Institute of Architects New Practices New York
Architectural League Young Architects + Designers

Studio Farris Architects

Studio Farris Architects is an architectural practice based in Antwerp, Belgium, founded by Italian architect Giuseppe Farris in 2008.

The studio's goal is to discover the intrinsic potential in every project, questioning the obvious, exploring the surroundings and cultural heritage. Architecture, interior architecture, furniture design, lighting and graphic design are all part of the same overall concept and are designed as a coherent whole, with particular attention to detail.

The work of Studio Farris Architects has been widely featured in international publications such as Architectural Record (USA), AZURE (Canada), TASARIM (Turkey) and Wallpaper* (United Kingdom). Studio Farris Architects is the recipient of Architectural Record's 2015 Design Vanguard Award.

www.studiofarris.com

SUPA architects schweitzer song

Ryul Song was born in Taejon, Korea, and graduated from Hong-Ik University Seoul and Technical University of Dortmund, Germany. Christian Schweitzer was born in Linz, Austria, and graduated from Kaiserslautern University of Technology, Germany. Both met during their master's degree studies at the Frankfurt Staedelschule Academy of Fine Arts and established in 2000 their studio SUPA architects schweitzer song in Frankfurt, Germany. In 2005 they branched out to Seoul, Korea, teaching at Korea National University of Arts, Seoul National University, Korea University, and Hanyang University.

SUPA is working within the narrow intersection of conceptual design, art, theory and education. They are focused on the conceptual approach towards architectural design, finding new ways to expand the vocabulary of architecture. Their special interest lies in the exploration of the specific sociocultural context inherent in a task as the interface between everyday life, art and architecture.

TAKK Architecture

Takk is a space for architectural production focused in the development of experimental and speculative material practices in the intersection between nature and culture in the contemporary framework, with a special attention on the overcoming of anthropocentrism on its different ways (political, ecological, cultural, on gender), and on the definition of new notions of beauty through the articulation of the difference by assembling a multiplicity of materials from different origins and conditions.

Additionally to this profesional practice, Takk is developing a framework in the field of research and teaching. Mireia Luzárraga and Alejandro Muiño are teachers in the Projects Department of the Universidad de Alicante (UA), in BAU Barcelona Design University Centre, and Master Tutors in the Institute of Advanced Architecture of Catalonia (IAAC). They have also participated as teachers in different workshops and summer schools and have explained their work in several lectures internationally.

At the present time, Mireia and Alejandro combine their profesional and teaching labour with the development of their respective PhD Thesis on the politics of ornament and self sufficient micro-communities. They have been granted for them with the scholarship "Junior Faculty - La Caixa".

TheeAe Architects

TheeAe is abbreviation of the evolved architectural eclectic. Its name is about the effort and dedication to the value of architectural aesthetic which shall be laid on place, history and culture of surrounding environment. TheeAe pursues re-searching and re-finding the elements that have been embedded into the context of environment, so as to define the beauty of the architecture within the given context.

In this passion, TheeAe has begun its practice, since 2011 in Hong Kong, to continue to explore and learn of culture and beauty of environment through the design. TheeAe's service has been extensively covered in various areas of architecture & interior design. The projects include, one of landmark projects, Mumbai Airport (Chhatrapati Shivaji International airport), and Mumbai Hotel (Bellagio & Mandarin) in India, Mardina Residence in Saudi Arabia. As well as, projects are developed and proposed in many other places around world, those of which are Afghanistan Museum in Kabul, Afghanistan, Guggenheim Museum in Helsinki, Finland, Science Center in Kaunas, Lithuania, Arch-lay Contemporary Museum in Shenzhen, China, Elevated Elf Land, horse theme park, in South Korea, Dubai Heart, Dubai, UAE., etc.

Our service will continue to serve the clients who are seeking for superior design quality not only to increase the value of the properties but also to bring the meanings of places for people, who will live, visit and pass by.

TOUCH Architect

TOUCH Architect Co.,Ltd. was first established since 2014. It was changed from TOUCH STUDIO Architect partnership, with four years experiences into a company. With a great chance of an improvement, we have two main co-founders consist of Mr. Setthakarn Yangderm as an architect and leader of our firm and Ms. Parpis Leelaniramol as an architect, which will corporate together in design, construction, and management.

SETTHAKARN YANGDERM
Architect // Managing Director // Founder
Bachelor of Architecture
(Department of Architecture, Major in Thai Architecture)
Faculty of Architecture, Chulalongkorn University. (1st top rank of Thailand's university)
Honor in best architectural design in 2007.

PARPIS LEELANIRAMOL
Architect // Co-founder
Bachelor of Science
(International Program in Design and Architecture),
Faculty of Architecture, Chulalongkorn University. (1st top rank of Thailand's university)
Master of Science in Real Estate Business (MRE),
Faculty of Commerce and Accountancy, Thammasat University. (1st top rank of business program)
Teaching Assistant - MRE Thammasat Personal Consultant - Mini-MRE at Ananda Development (Real Estate Listed Company in Thailand)

UNStudio

© Inga Powilleit

Ben van Berkel studied architecture at the Rietveld Academy in Amsterdam and at the Architectural Association in London, receiving the AA Diploma with Honours in 1987.

In 1988 he and Caroline Bos set up an architectural practice in Amsterdam, extending their theoretical and writing projects to the practice of architecture. UNStudio presents itself as a network of specialists in architecture, urban development and infrastructure. Current projects include the design for Doha's Integrated Metro Network in Qatar, 'Four' a large-scale mixed-use project in Frankfurt and the Wasl Tower in Dubai.

With UNStudio he realised amongst others the Mercedes-Benz Museum in Stuttgart, Arnhem central Station in the Netherlands, the Raffles City mixed-use development in Hangzhou, the Canaletto Tower in London, a private villa up-state New York and the Singapore University of Technology and Design.

In 2018 Ben van Berkel founded UNSense, an Arch Tech company that designs and integrates human-centric tech solutions for the built environment.

Ben van Berkel has lectured and taught at many architectural schools around the world. Currently he holds the Kenzo Tange Visiting Professor's Chair at Harvard University Graduate School of Design, where he has led a studio on health and architecture. In 2017, Ben van Berkel also gave a TEDx presentation about health and architecture. In addition, he is a member of the Taskforce Team / Advisory Board Construction Industry for the Dutch Ministry of Economic Affairs.

A case study of

Glă

Architecture

Glass House, Granada, Spain

OFIS arhitekti

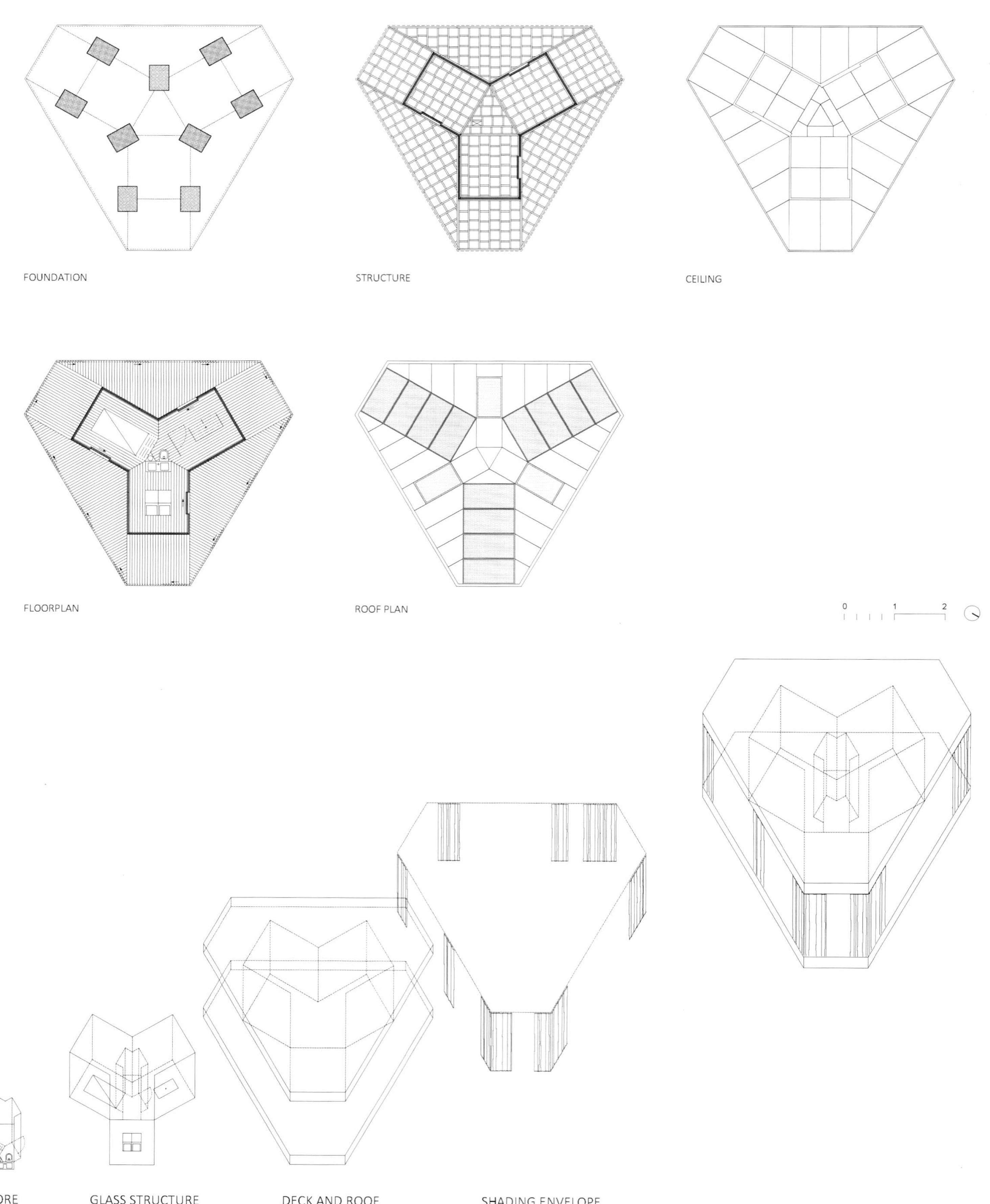

Concept Diagram

© Jose Navarrete

© Jose Navarrete

© Jose Navarrete

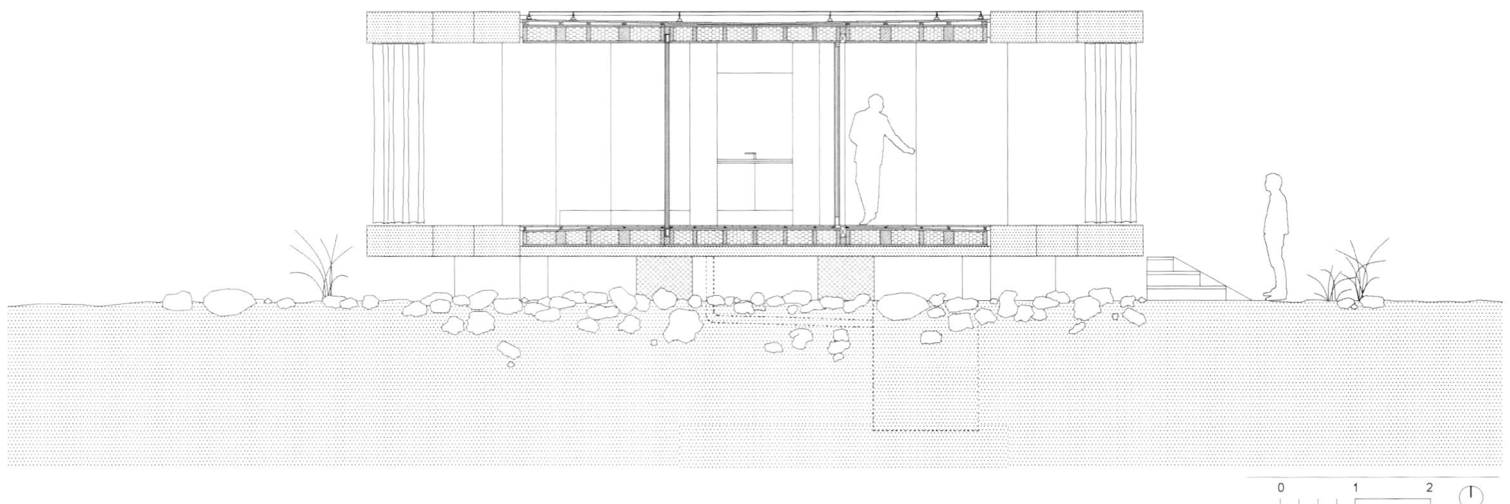

Cross Section

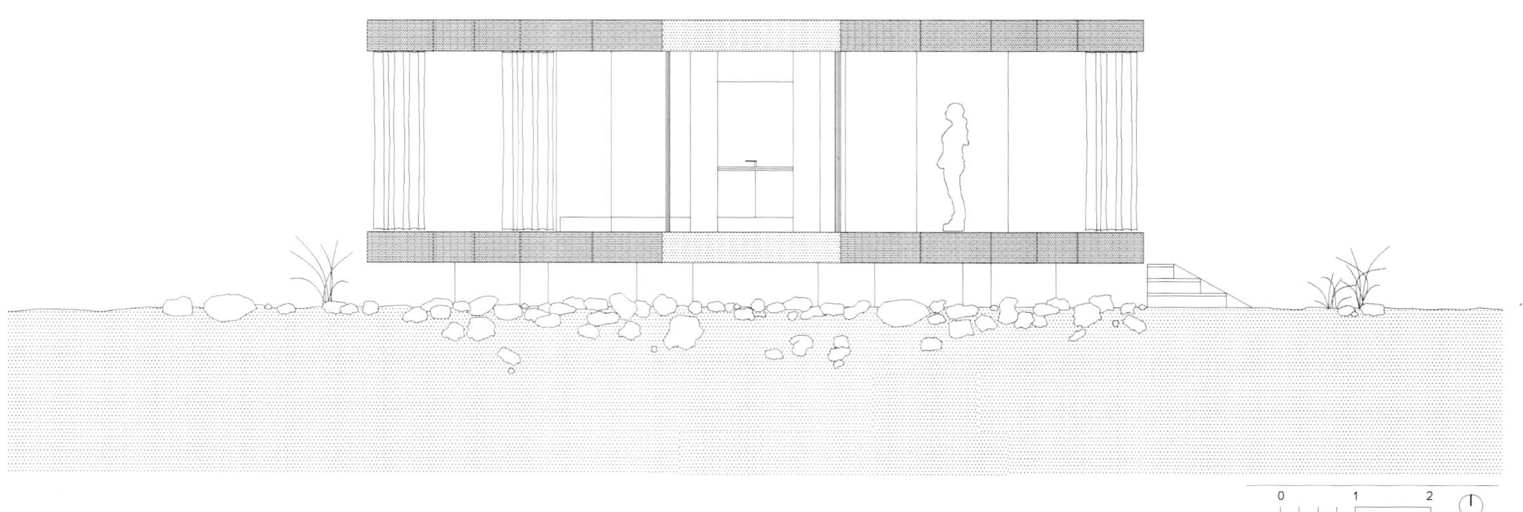

North Elevation

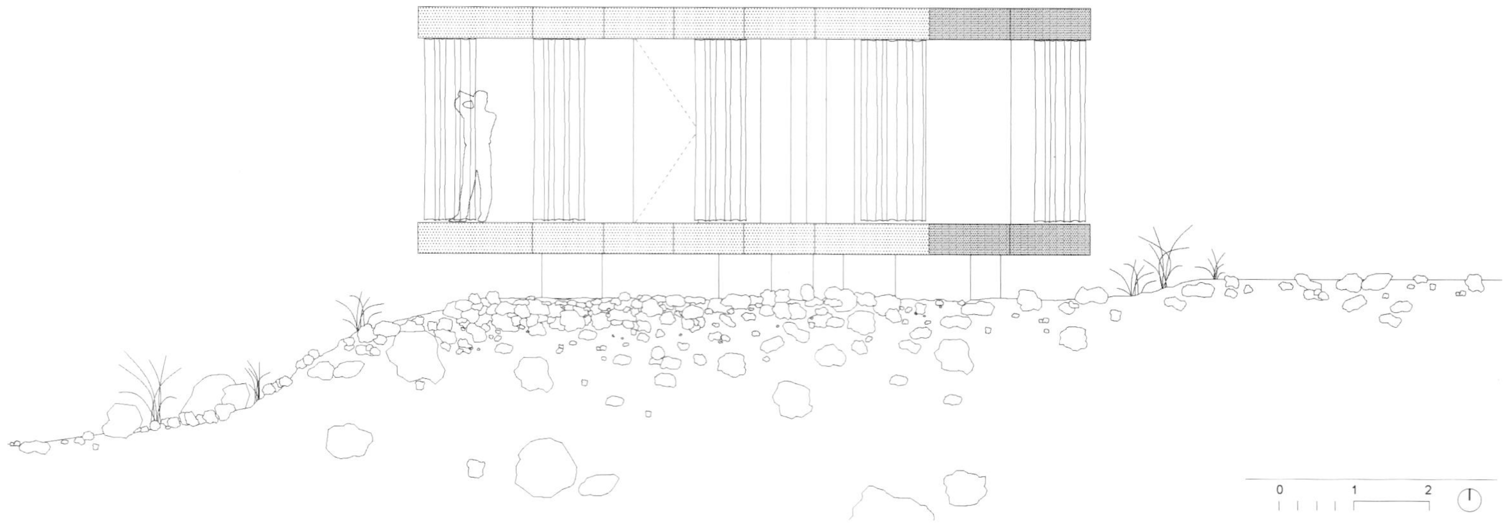

East Elevation

Construction Process

© Jose Navarrete

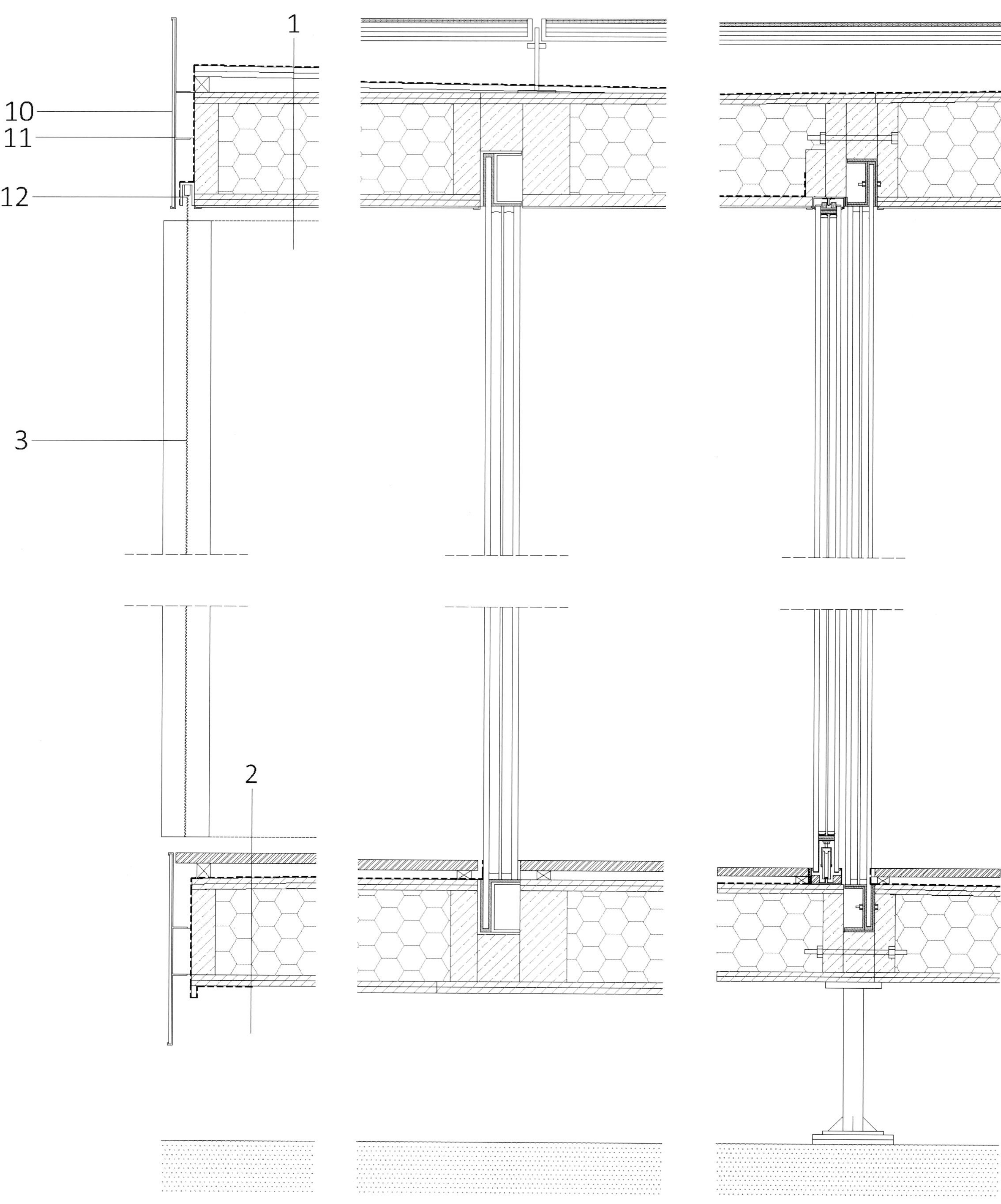

Construction Detail

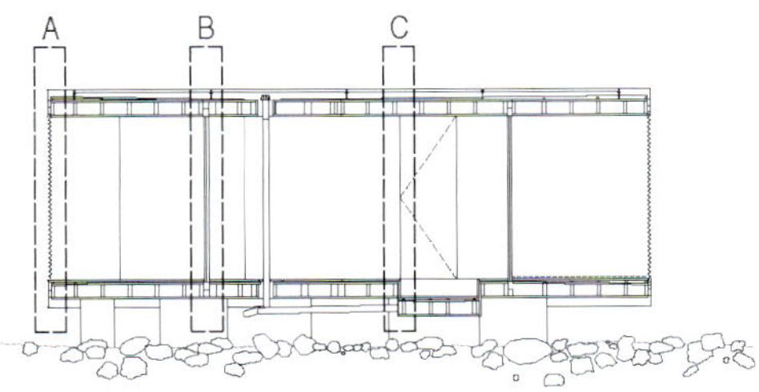

1 ROOF CONSTRUCTION
- PV PANELS 50mm
- PV PANELS SUBSTRUCTURE
- WATERPROOFING
- STRUCTURAL PLYWOOD 2x12mm
- C24 TIMBER JOIST 50x200mm
- THERMAL INSULATION 200mm
- REMOVABLE SOFIT 24mm
- VAPOR BARRIER
- GLASS CLADDING 4mm

2 FLOOR CONSTRUCTION
- OAK BOARD FLOORING 22mm
- WATERPROOFING
- STRUCTURAL PLYWOOD 2x12mm
- THERMAL INSULATION 200mm
- STRUCTURAL PLYWOOD 2x12mm

3 SYNTHETIC CURTAIN
4 GLASS ENVELOPE 62,5mm
5 BATHTUB
6 SLIDING DOOR
7 PVC ROOF DRAIN
8 PVC DRAIN PIPE Ø90mm
9 GLASS PARTITION 10mm
10 GLASS CLADDING
11 METALLIC SUBSTRUCTURE
12 CURTAIN RAIL

1 2

© Jose Navarrete

© Jose Navarrete

© Jose Navarrete

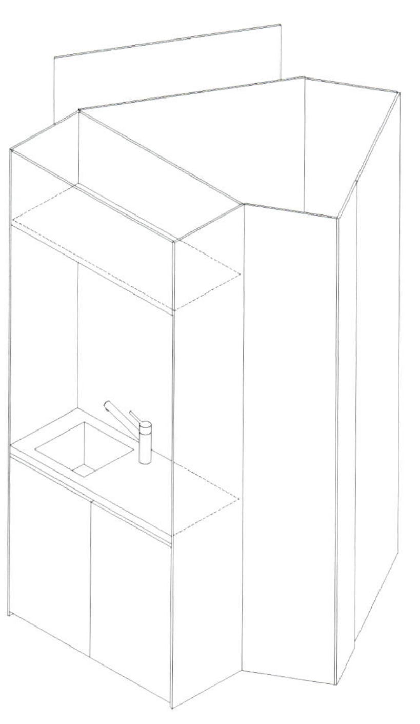

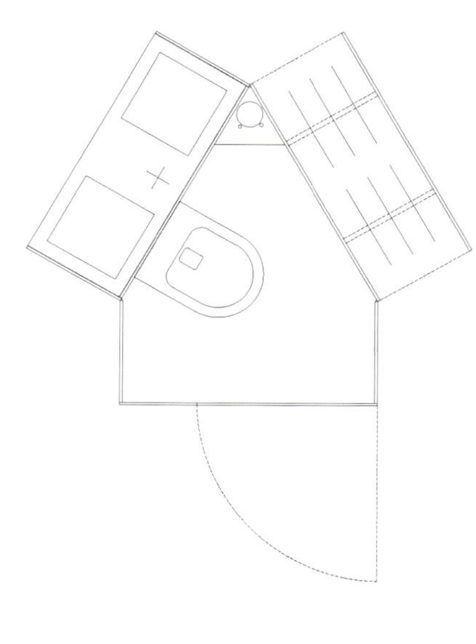

Construction Detail

© Jose Navarrete

© Jose Navarrete

Izmir Chamber Of Geological Engineers Building, Zmir, Turkey

M artı D Mimarlık

© ZM Yasa Photography

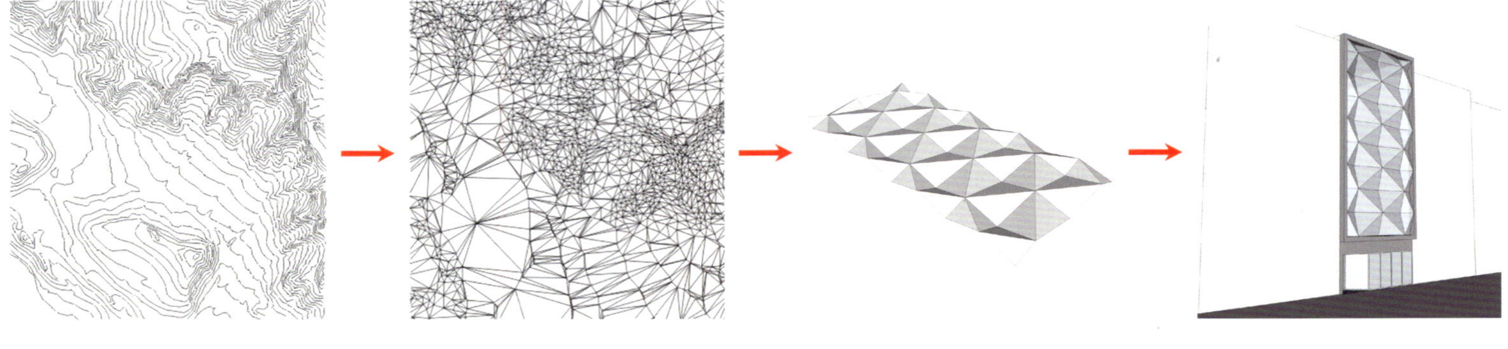

elevation curves

topographical triangulation

triangulation based facade scheme

triangulation based facade scheme

Facade Design Diagram

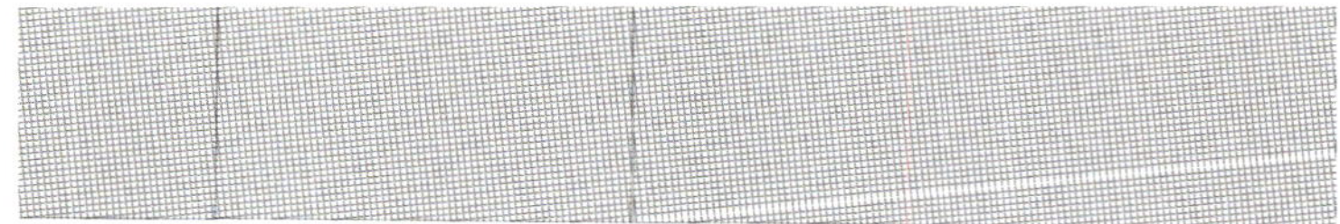

Ground Floor

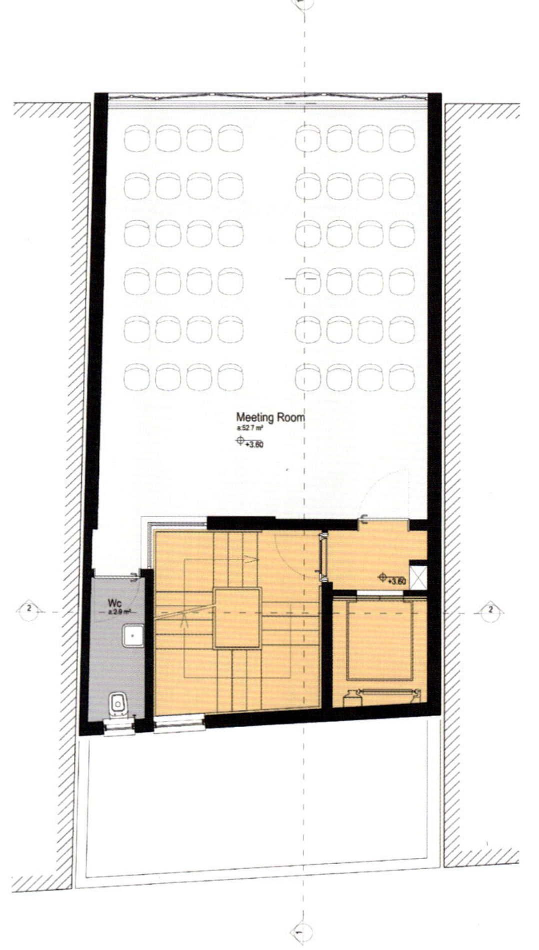

First Floor

© ZM Yasa Photography

© ZM Yasa Photography

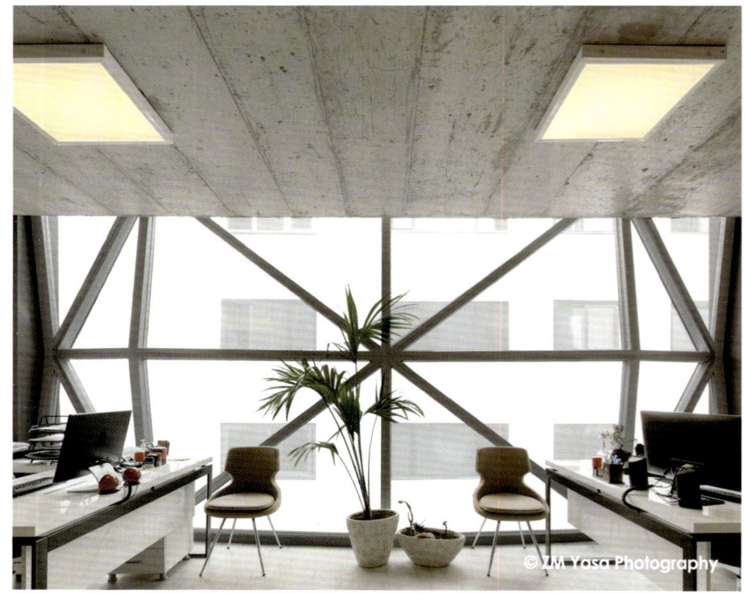

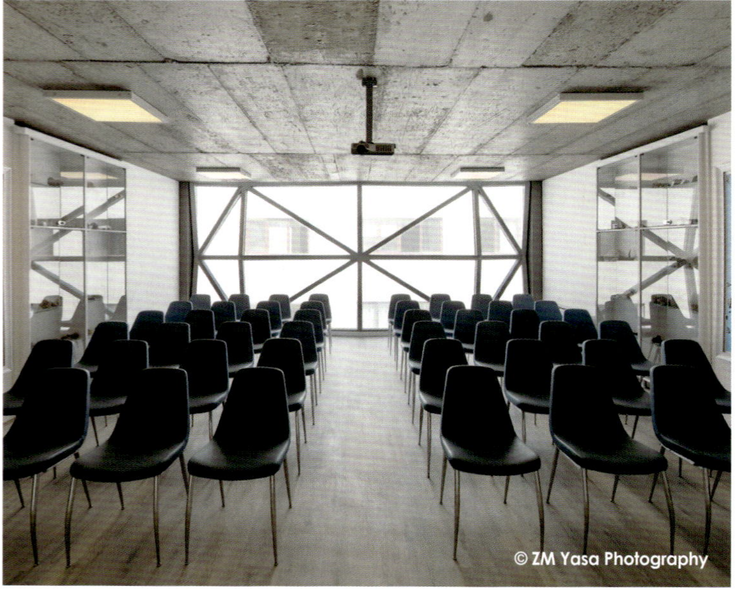

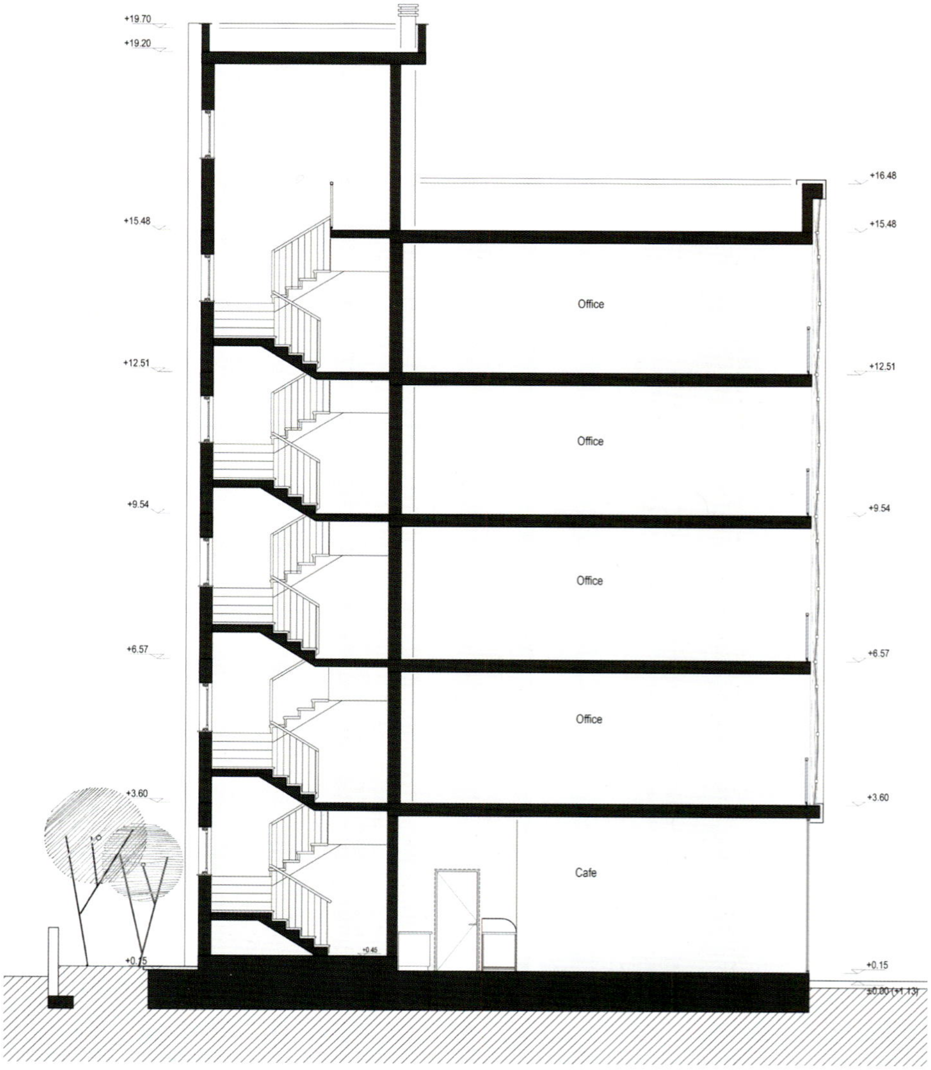

+19.70
+19.20
+16.48
+15.48
+15.48
Office
+12.51
+12.51
Office
+9.54
+9.54
Office
+6.57
+6.57
Office
+3.60
+3.60
Cafe
+0.45
+0.15
+0.15
±0.00 (+1.13)

Section 1-1

+19.20
+16.98
+15.48
Wc
+12.51
Wc
+9.54
Wc
+6.57
Wc
+3.60
Hol
+0.45
+0.15
±0.00 (+1.13)

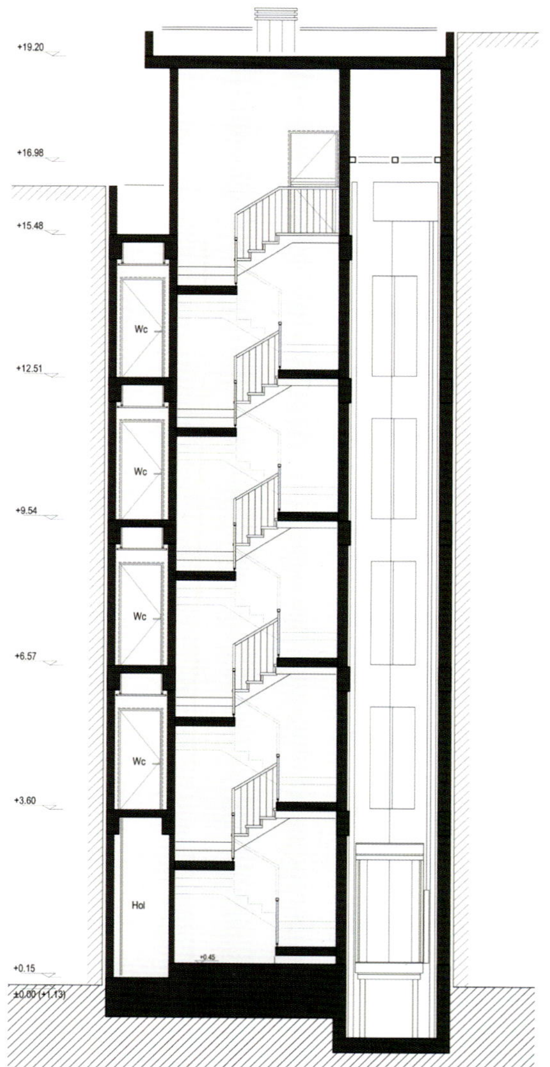

Section 2-2

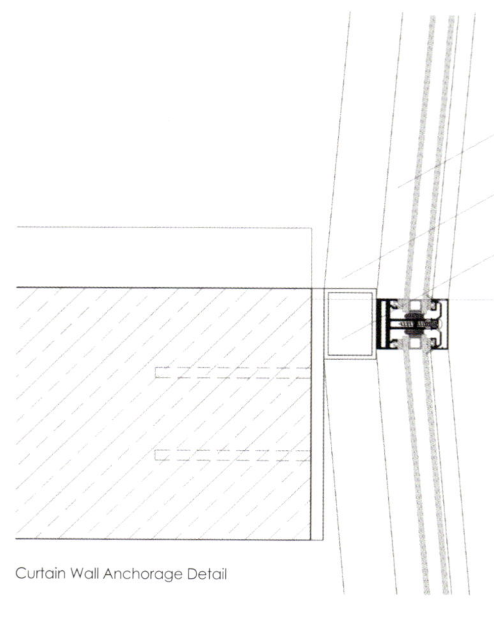

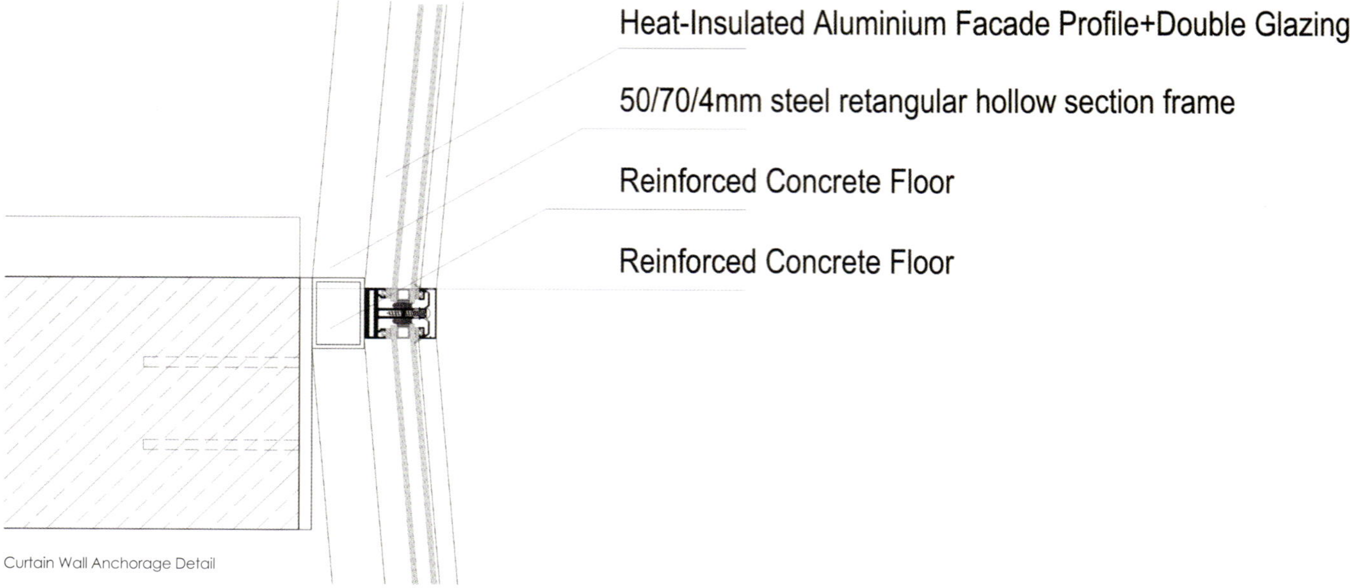

Curtain Wall Anchorage Detail

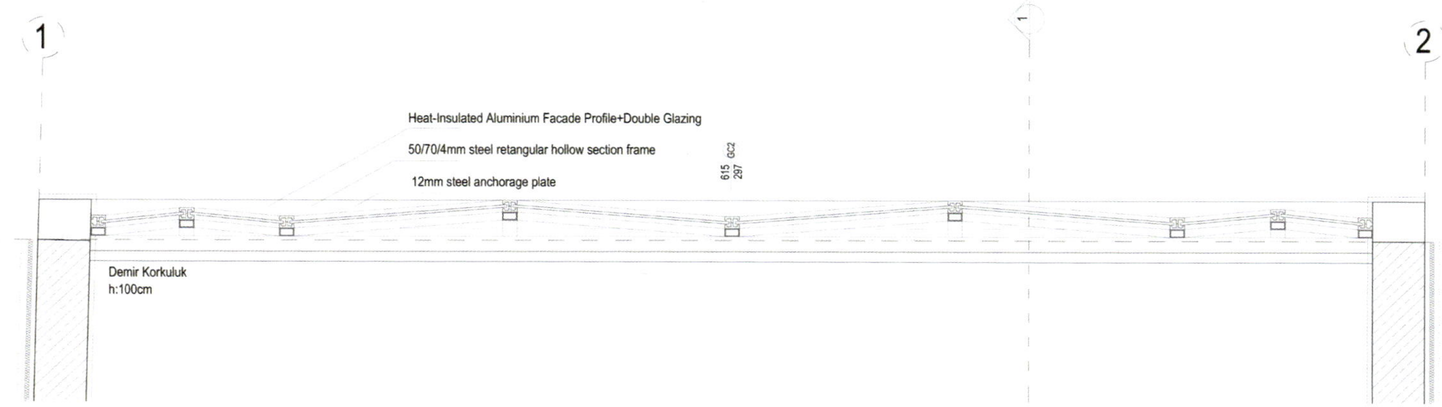

Curtain Wall Plan

© ZM Yasa Photography

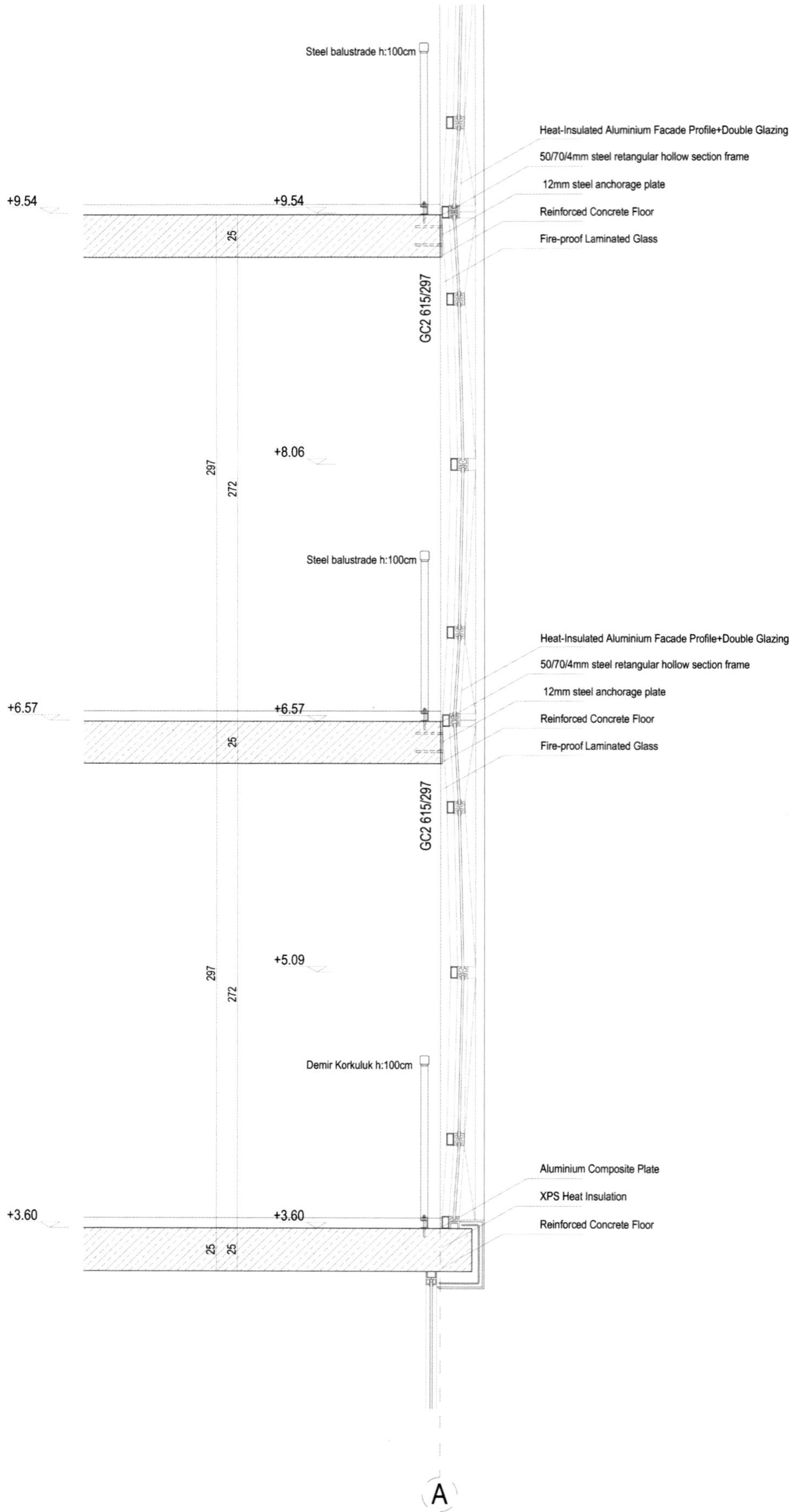

Curtain Wall System Section Detail

© ZM Yasa Photography

Au Pont Rouge, Saint Petersburg, Russia

Cheungvogl

Pub Oldham
PUB

МУЖСКОЕ БѢЛЬЕ
ЮБКИ
DEUIL LIVREE
FOURRURE
ЭСДЕРСЪ и СХЕФАЛЬСЪ

AU PONT ROUGE
ESDERS ET SCHEEFHALS
ТУЖУРКИ
ОБУВЬ
МѢХОВЫЯ ВЕЩИ
ШЛЯПЫ
AU PONT ROUGE
LINGERIE
CHAPEAUX
GANTERIE
CHAUSSURES
ЗИНЪ У КРАСНАГО МОСТА
ЭСДЕРСЪ и СХЕФАЛЬСЪ

BEN THANH restaurant, Ho Chi Minh City, Viet Nam
NISHIZAWA ARCHITECTS

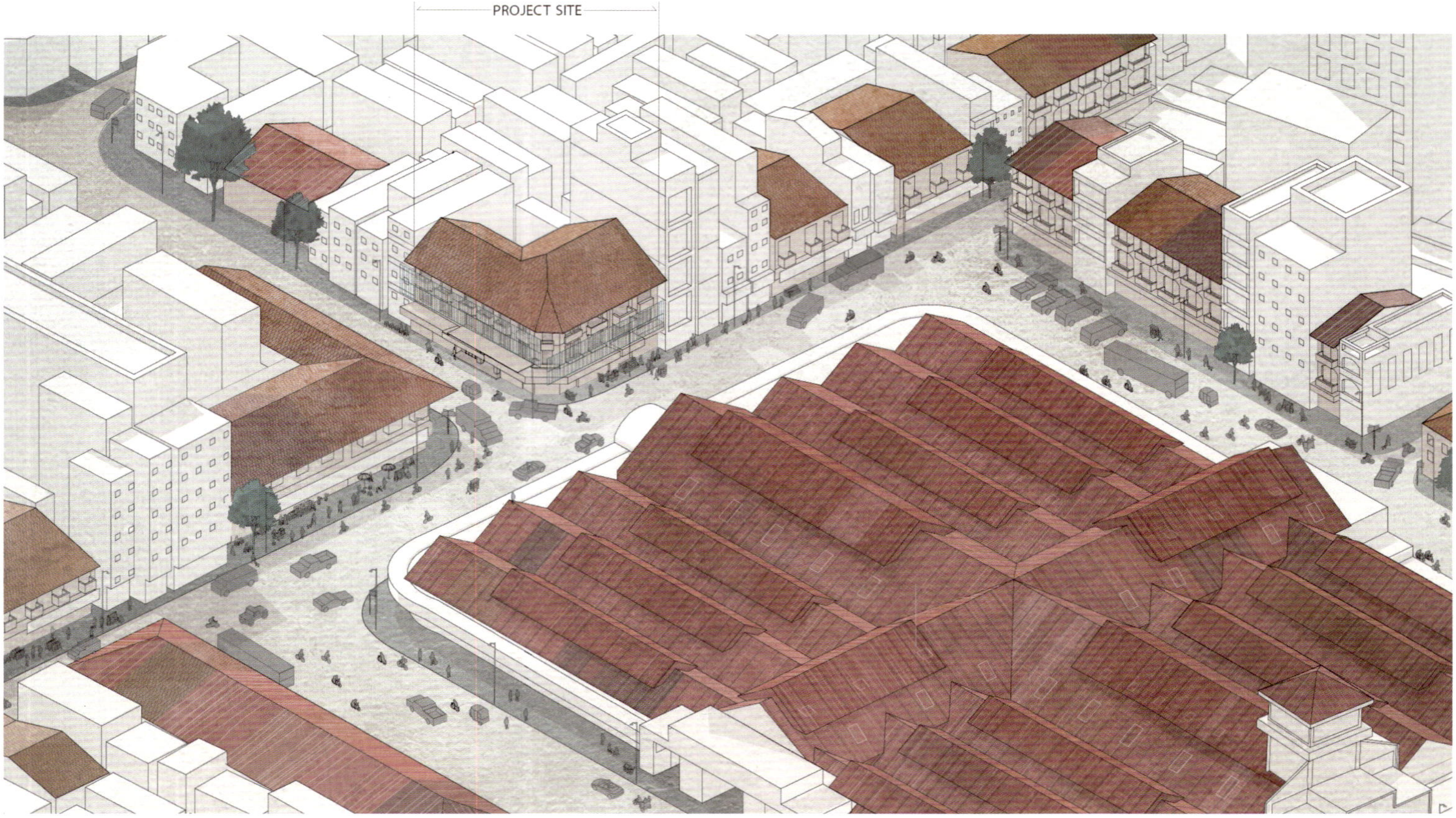

PROJECT SITE

Site Plan

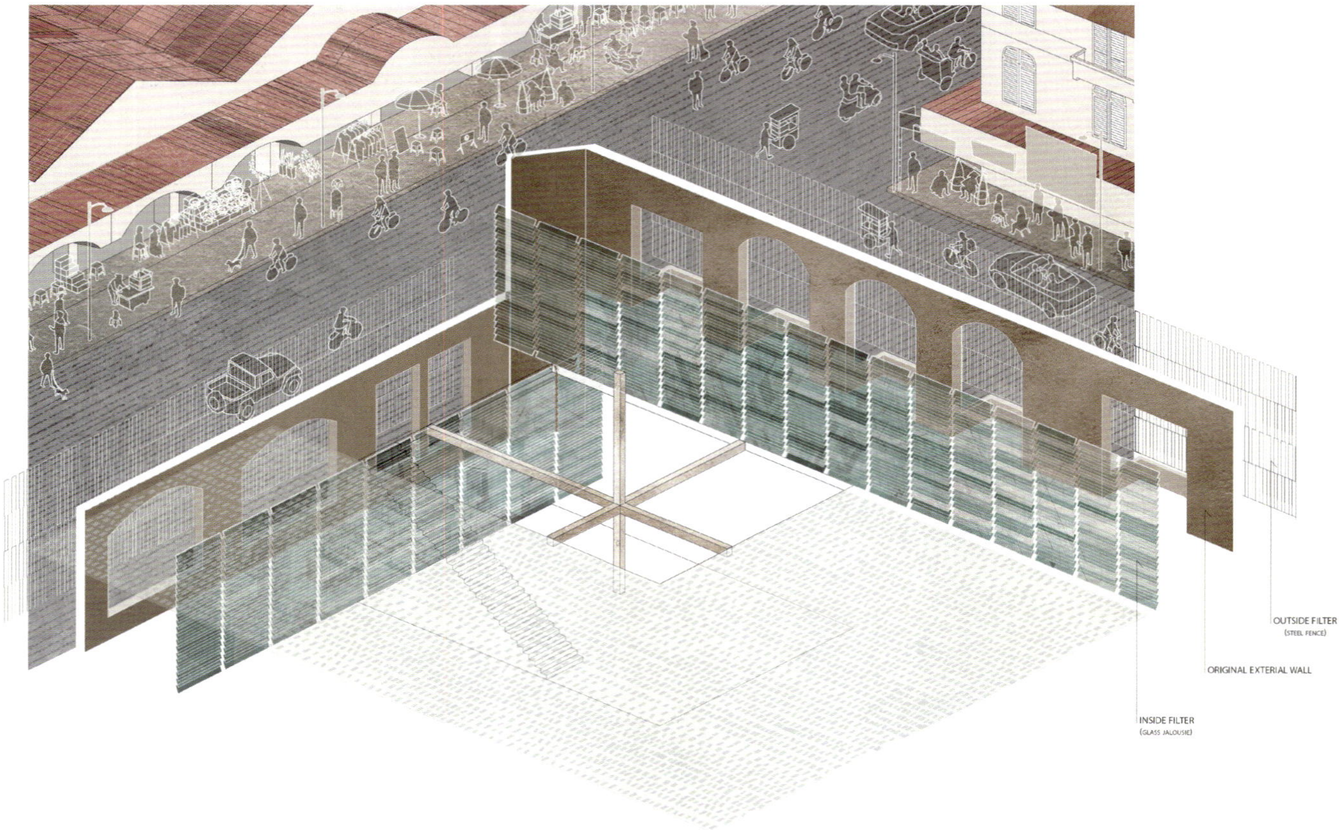

Detail Axonometirc

© NISHIZAWA ARCHITECTS

© NISHIZAWA ARCHITECTS

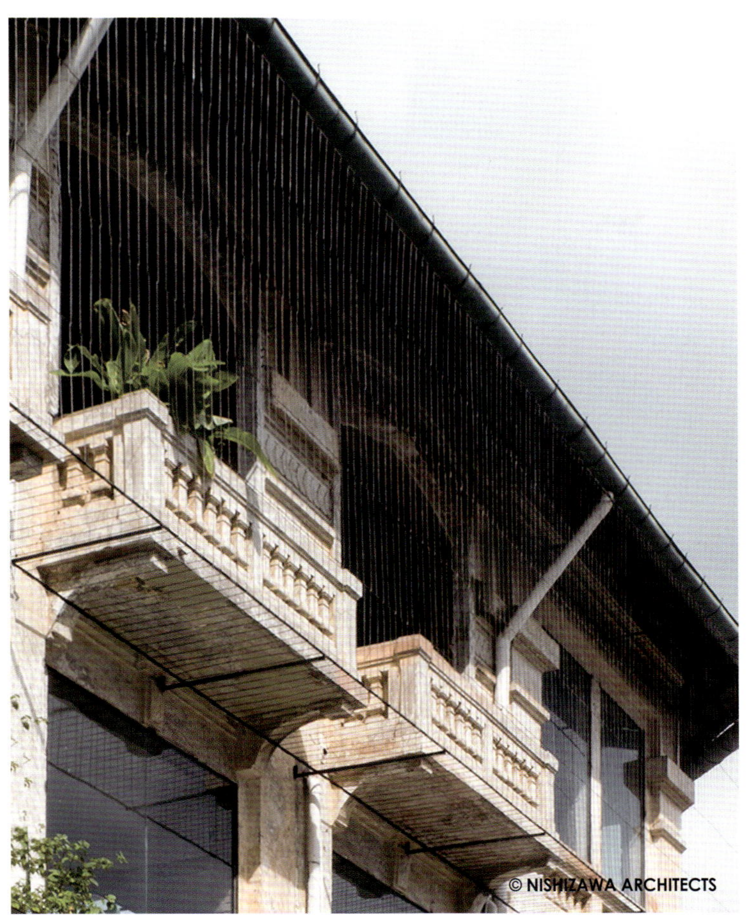

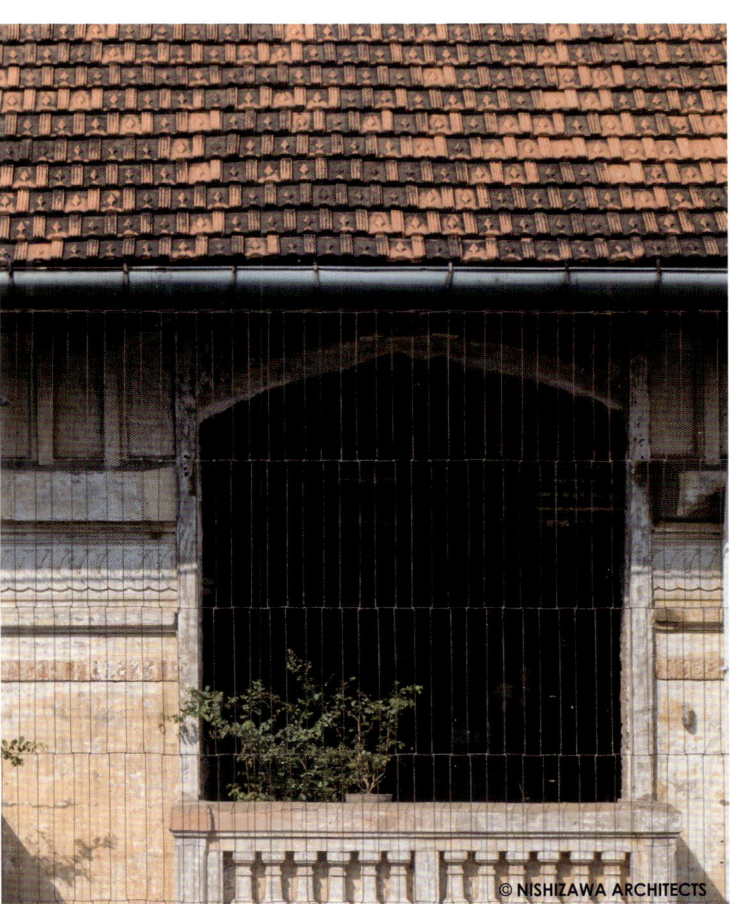

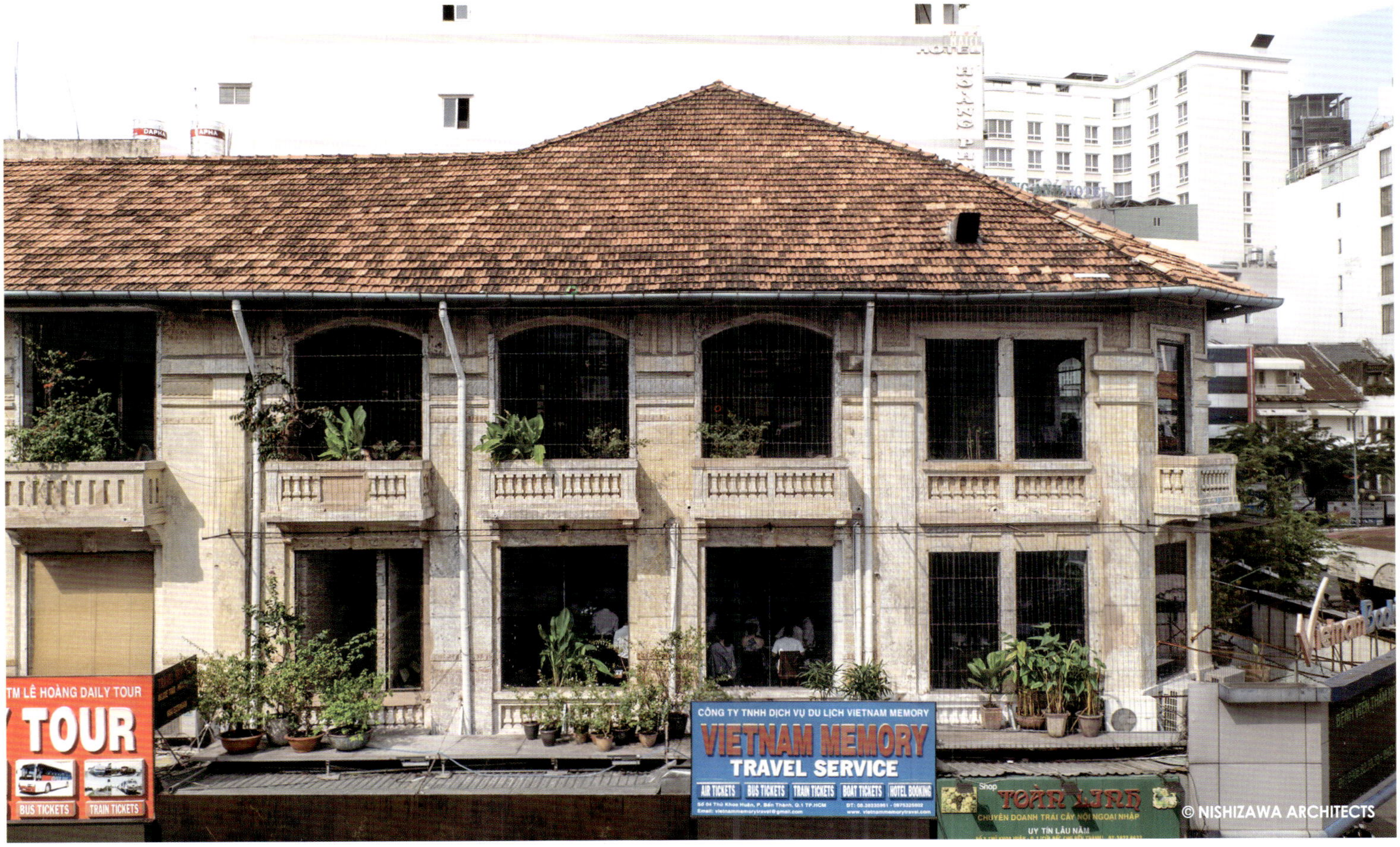
TM LÊ HOÀNG DAILY TOUR
TOUR
BUS TICKETS
TRAIN TICKETS
CÔNG TY TNHH DỊCH VỤ DU LỊCH VIETNAM MEMORY
VIETNAM MEMORY
TRAVEL SERVICE
AIR TICKETS
BUS TICKETS
TRAIN TICKETS
BOAT TICKETS
HOTEL BOOKING
TOÀN LINH
© NISHIZAWA ARCHITECTS

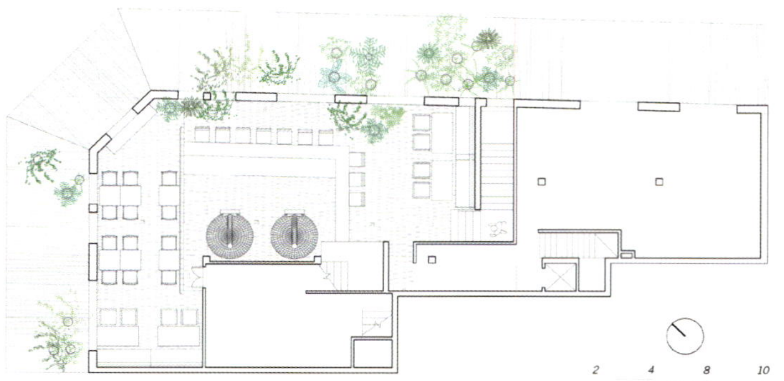

First Floor

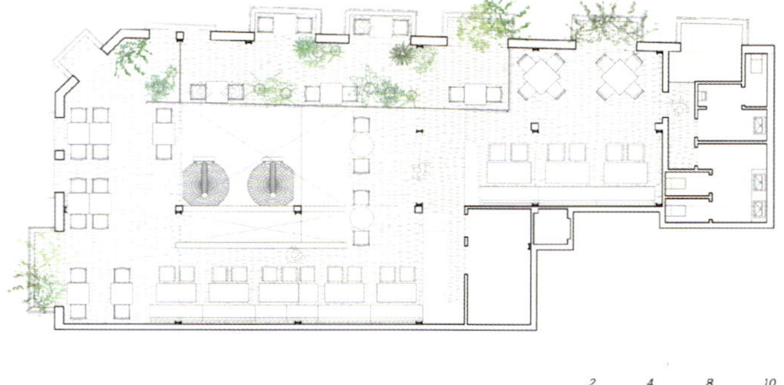

Second Floor

2 4 8 10

Longitude Section

© Hiroyuki Oki

© Hiroyuki Oki

© Hiroyuki Oki

© Hiroyuki Oki

© Hiroyuki Oki

© NISHIZAWA ARCHITECTS

Aviator, NY, USA

Urban Office Architecture(UOA)

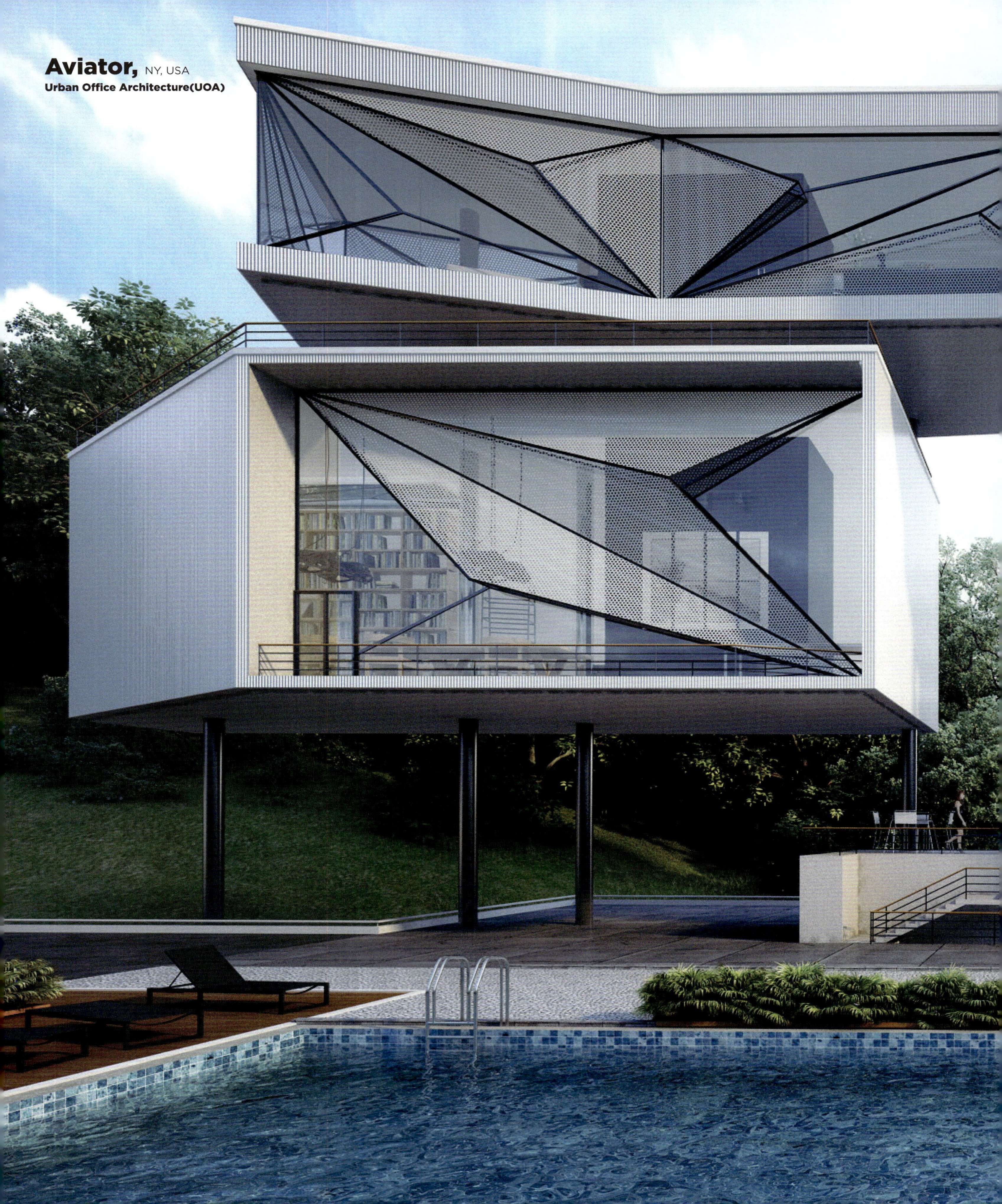

Site Plan

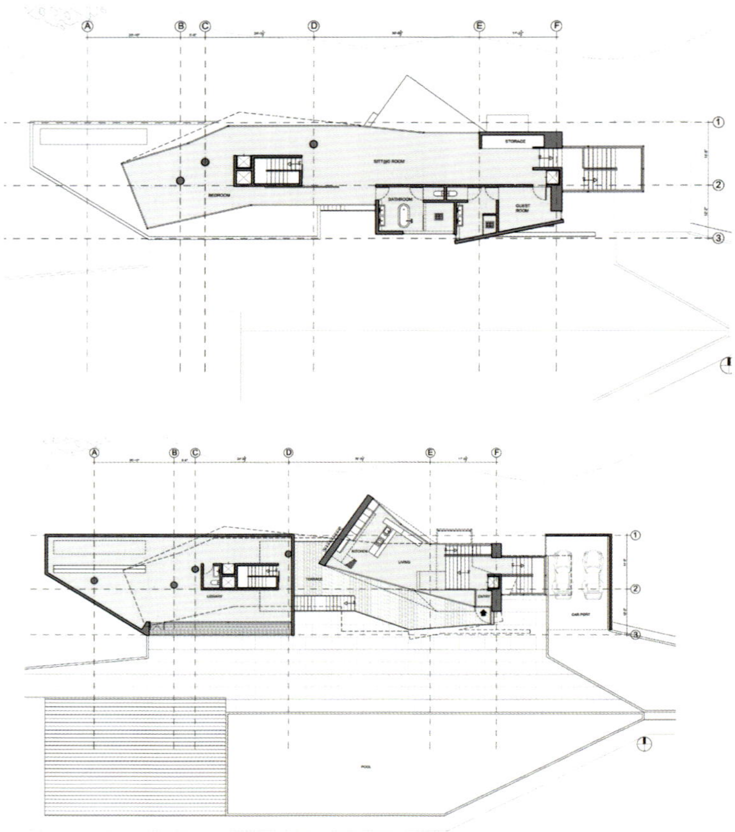

Floor Plan

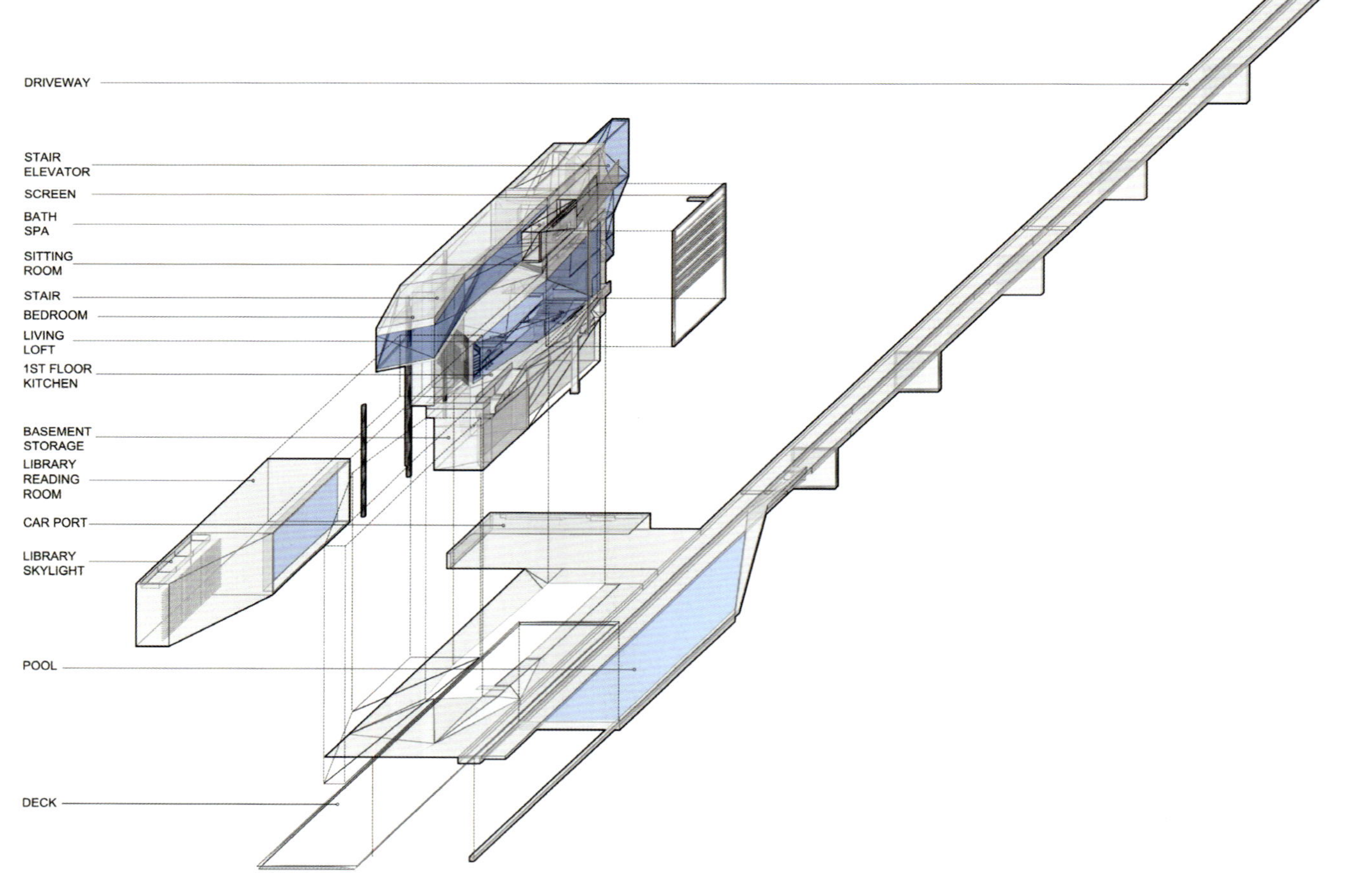

Axonometric

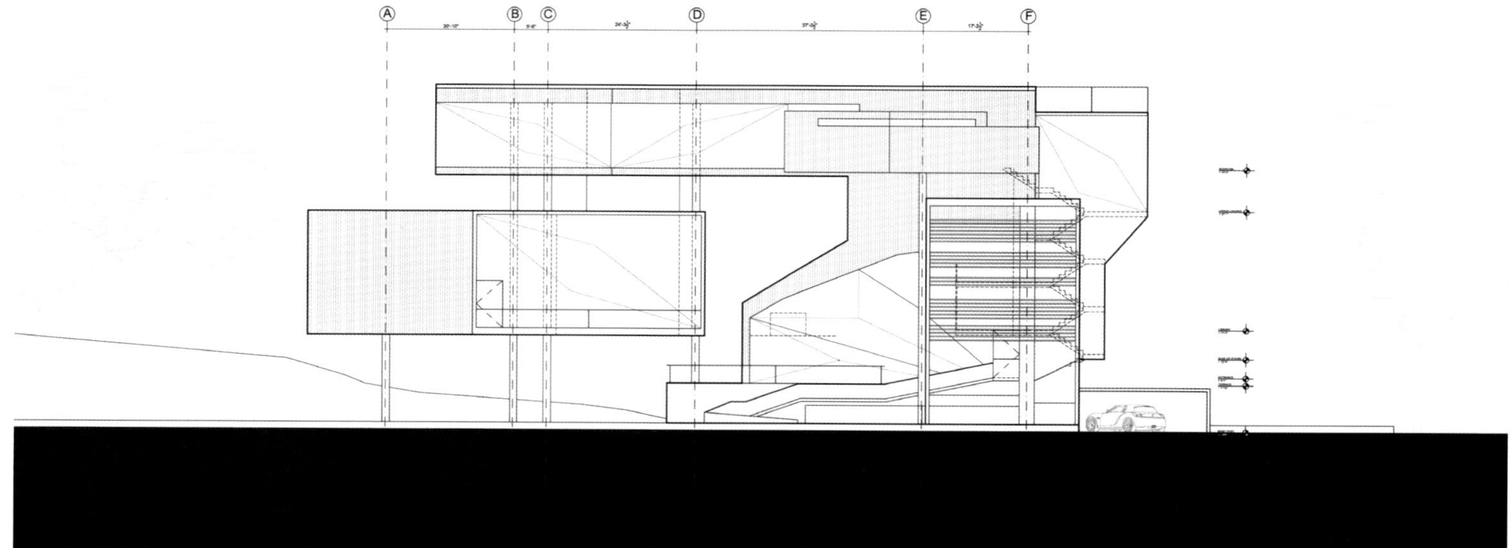

Elevation

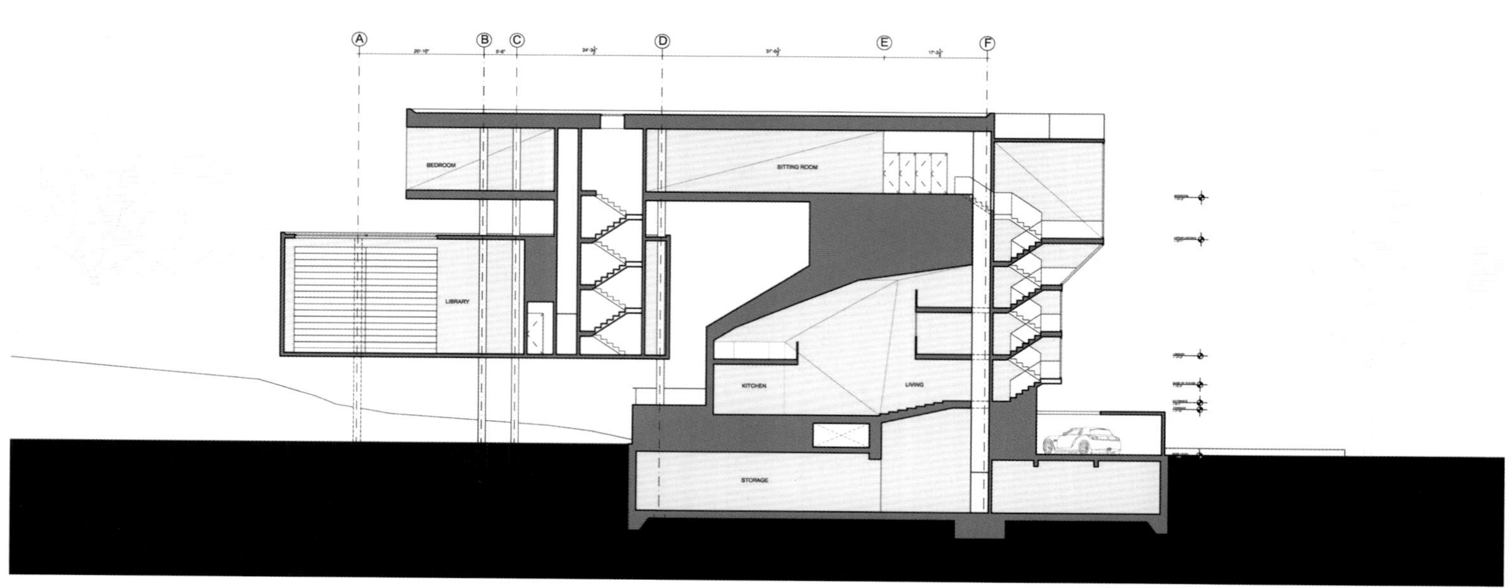

Section

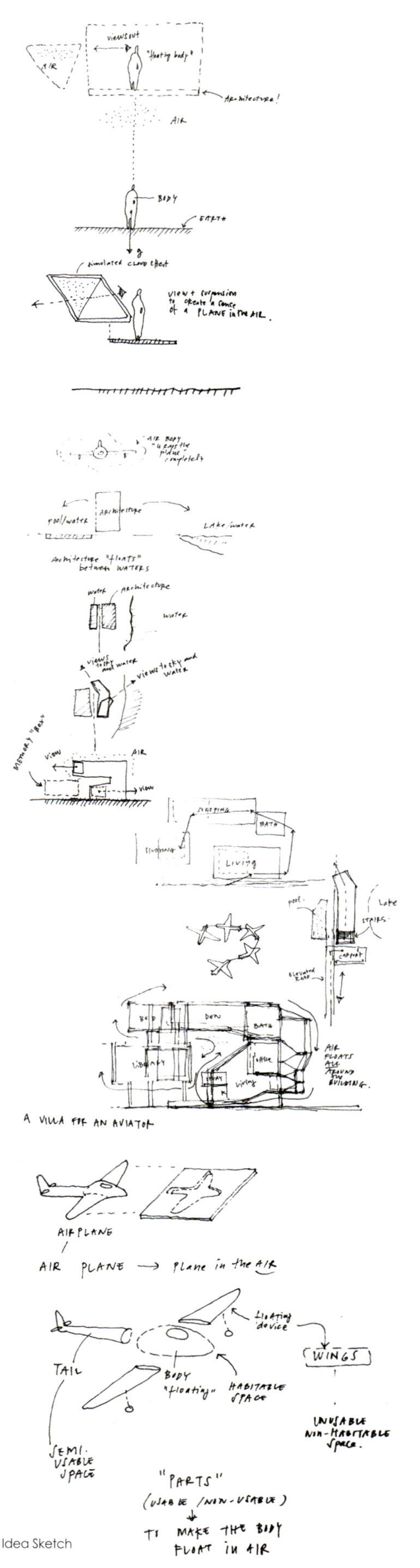
AIR
BODY
EARTH
A VILLA FOR AN AVIATOR
AIRPLANE
AIR PLANE → plane in the AIR
floating device
WINGS
TAIL
BODY "floating"
HABITABLE SPACE
UNUSABLE NON-HABITABLE SPACE
SEMI-USABLE SPACE
"PARTS"
(USABLE / NON-USABLE)
TO MAKE THE BODY FLOAT IN AIR

Idea Sketch

Cambeses House, Barcelos, Portugal

Rui Grazina

© Nelson Garrido

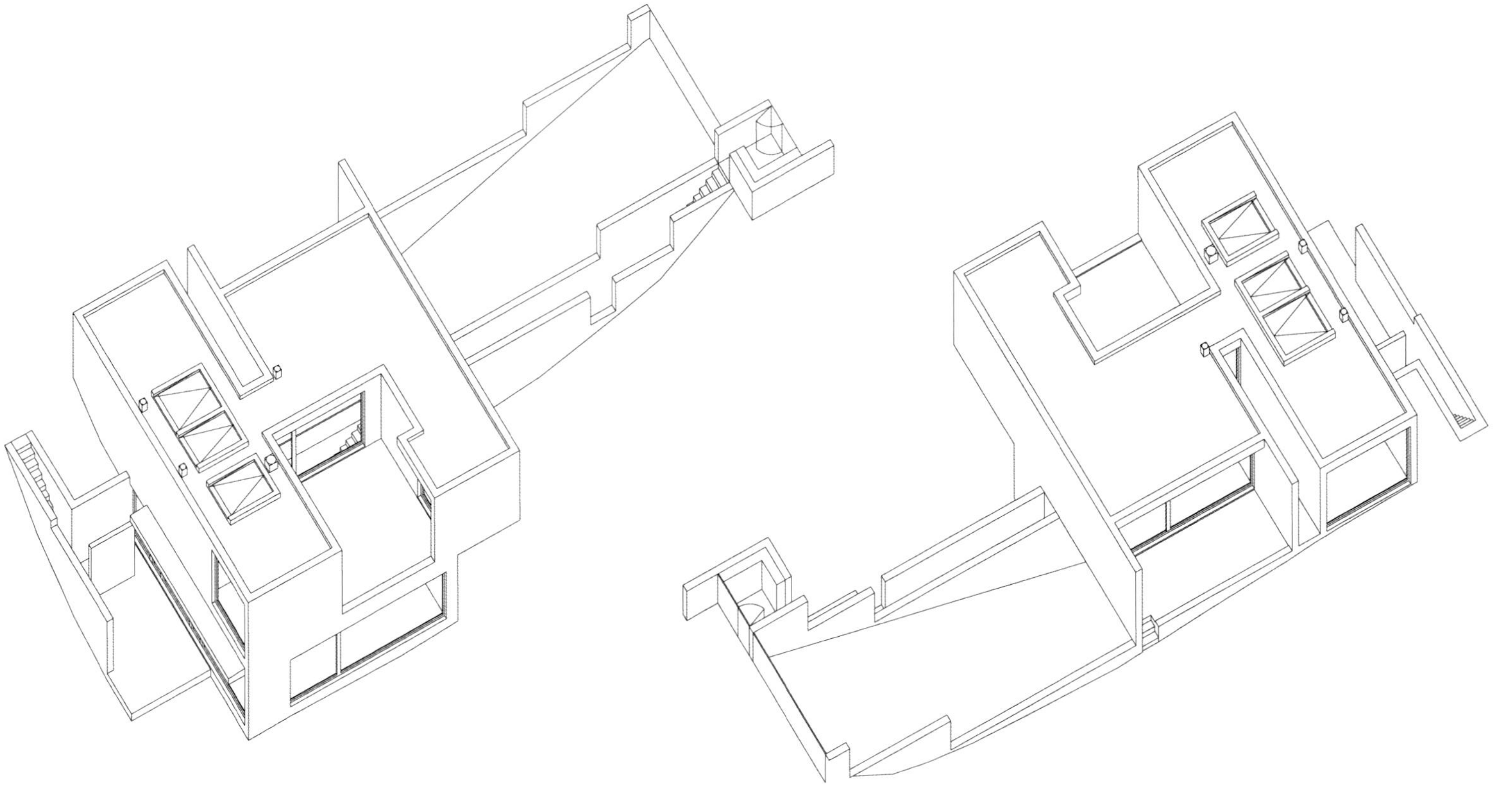

Axonometric

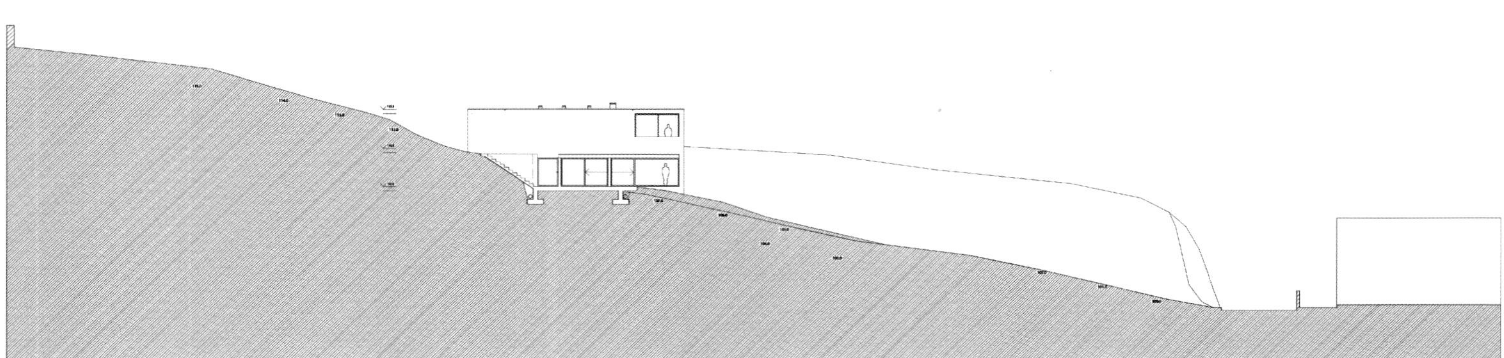

Elevation

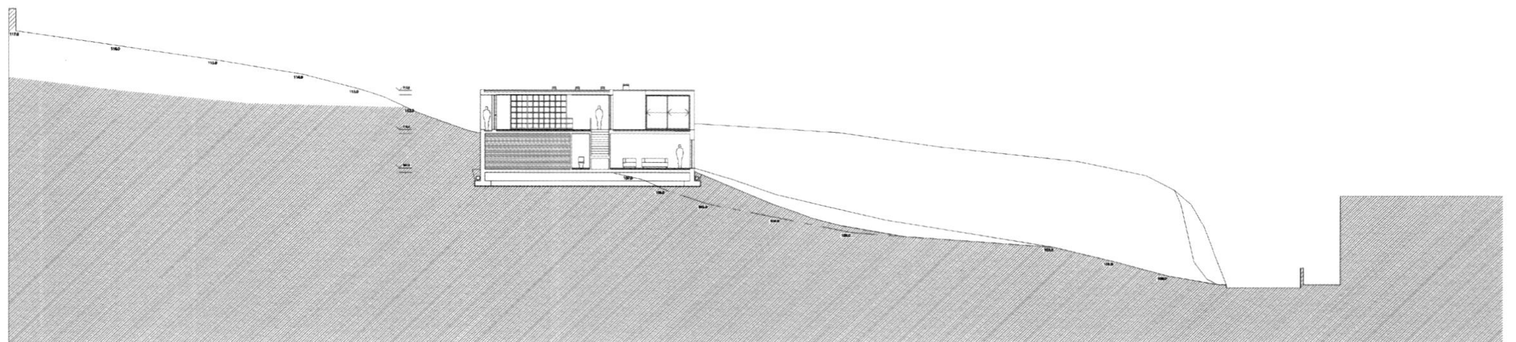

Section

© Nelson Garrido

© Nelson Garrido

© Nelson Garrido

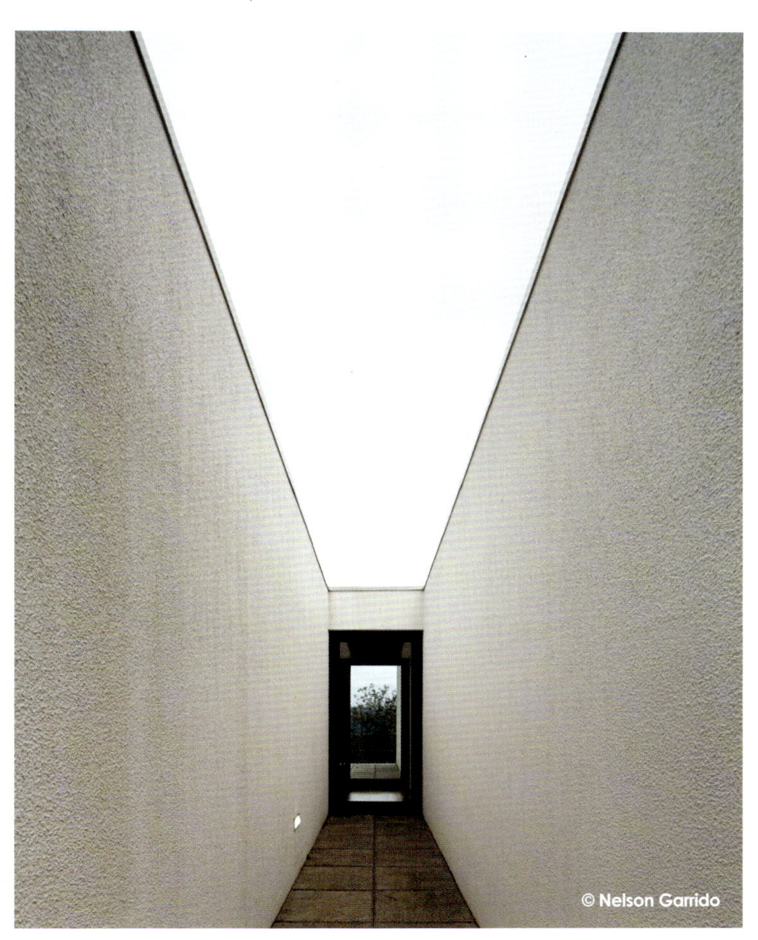

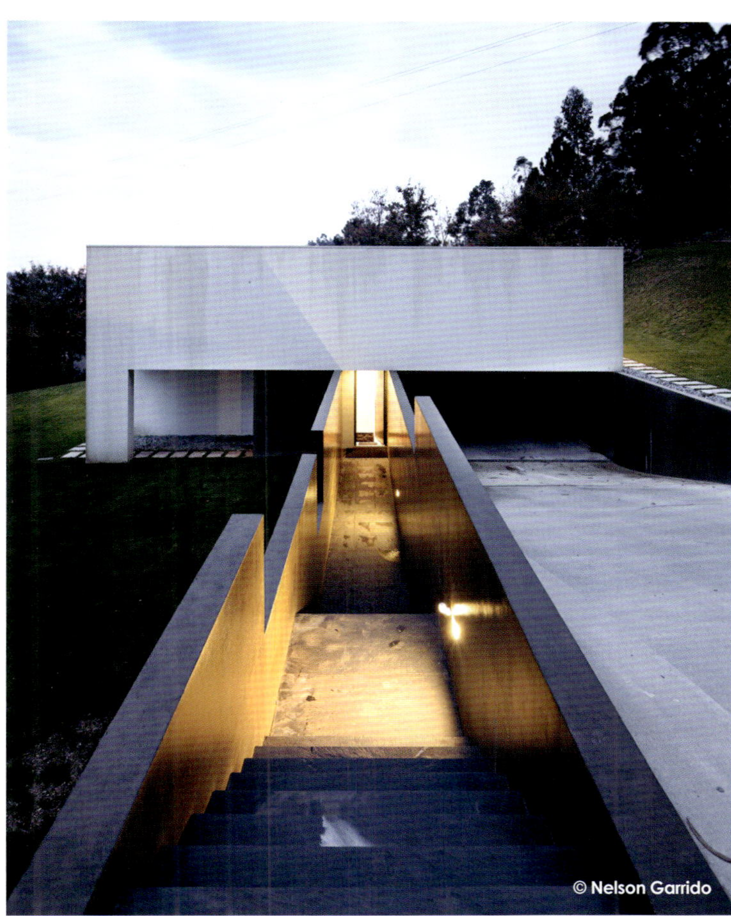

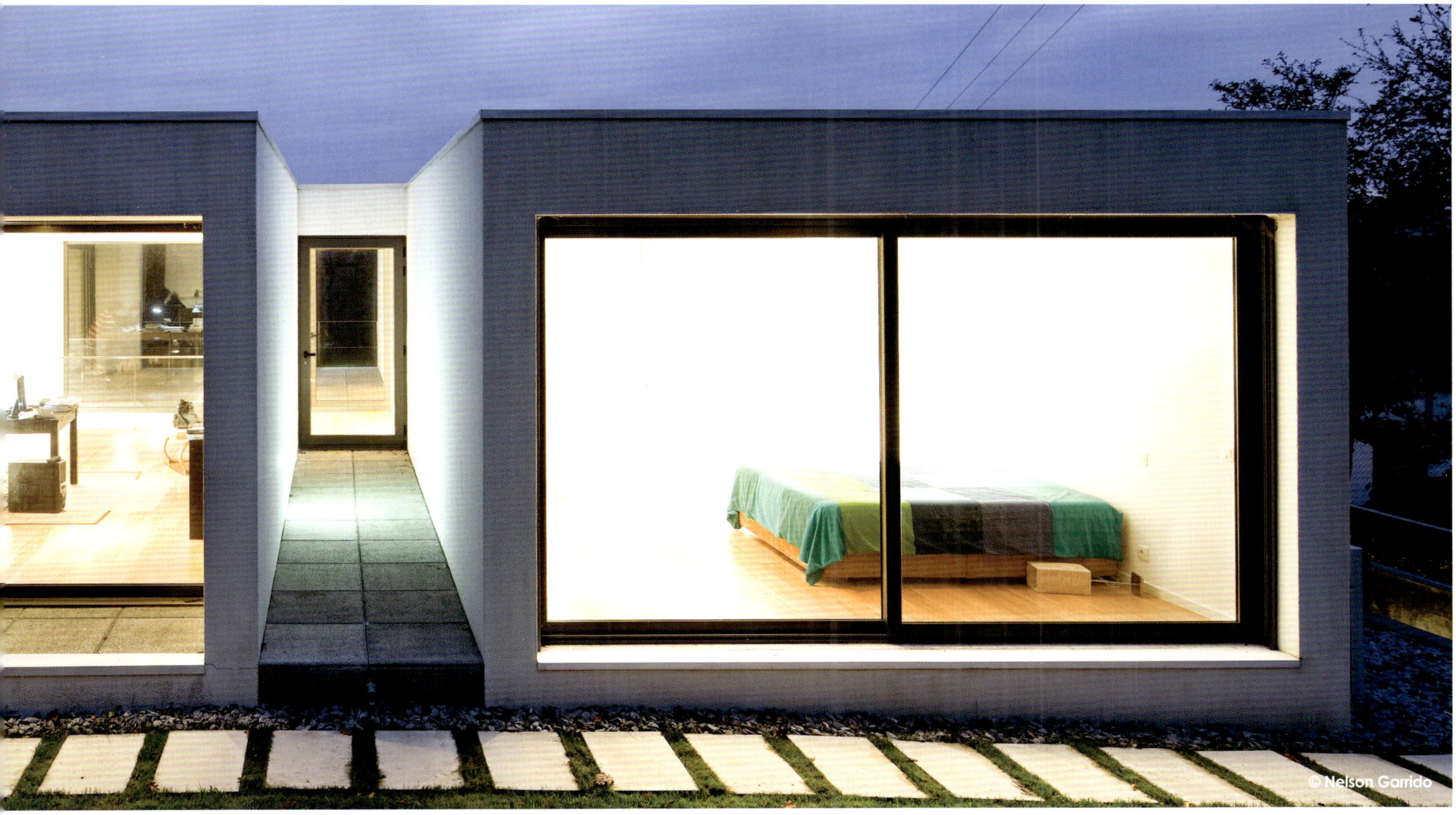
© Nelson Garrido

HAUS+, Hameln, Germany
Anne Menke

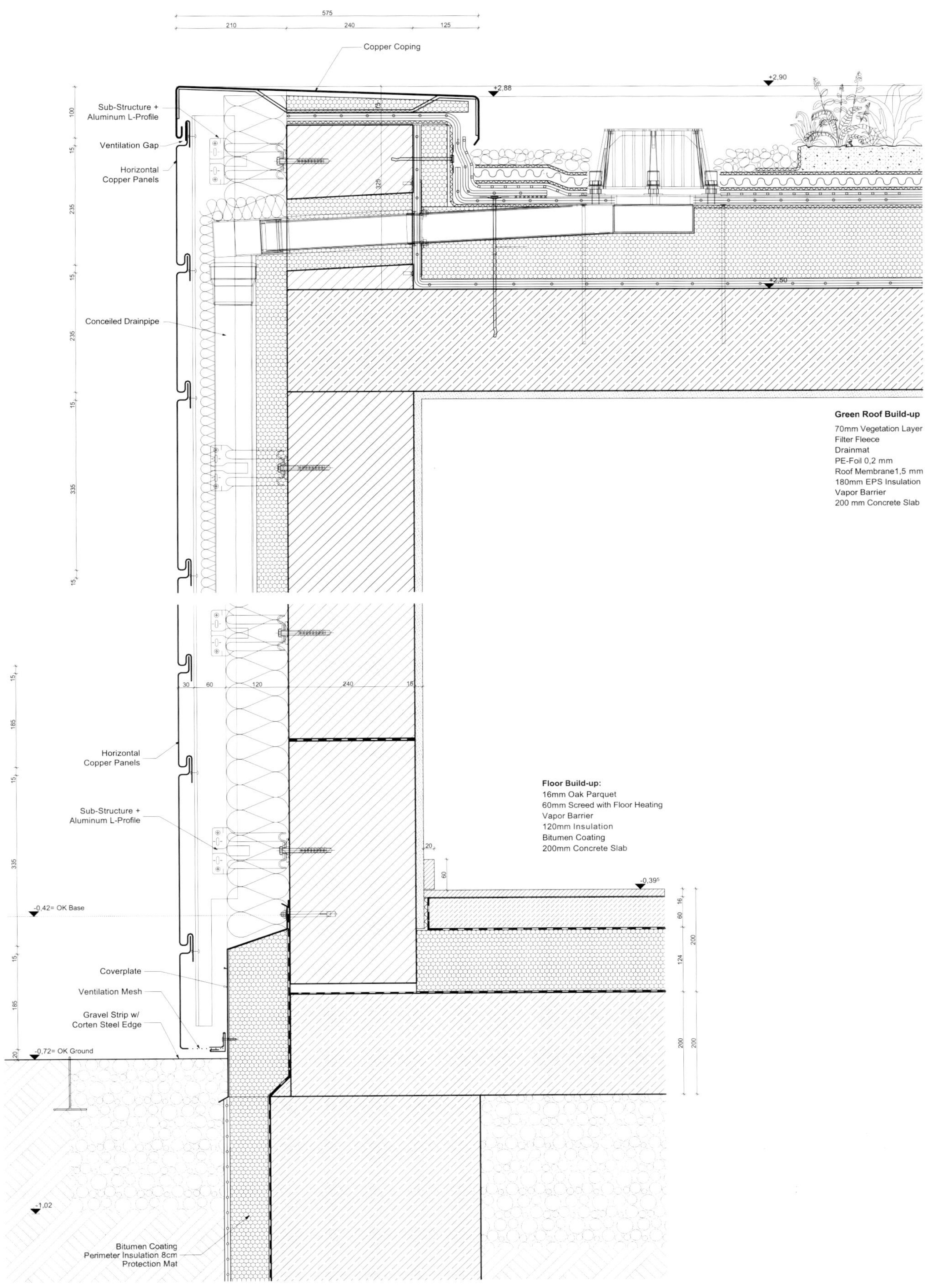

Copper Facade Detail

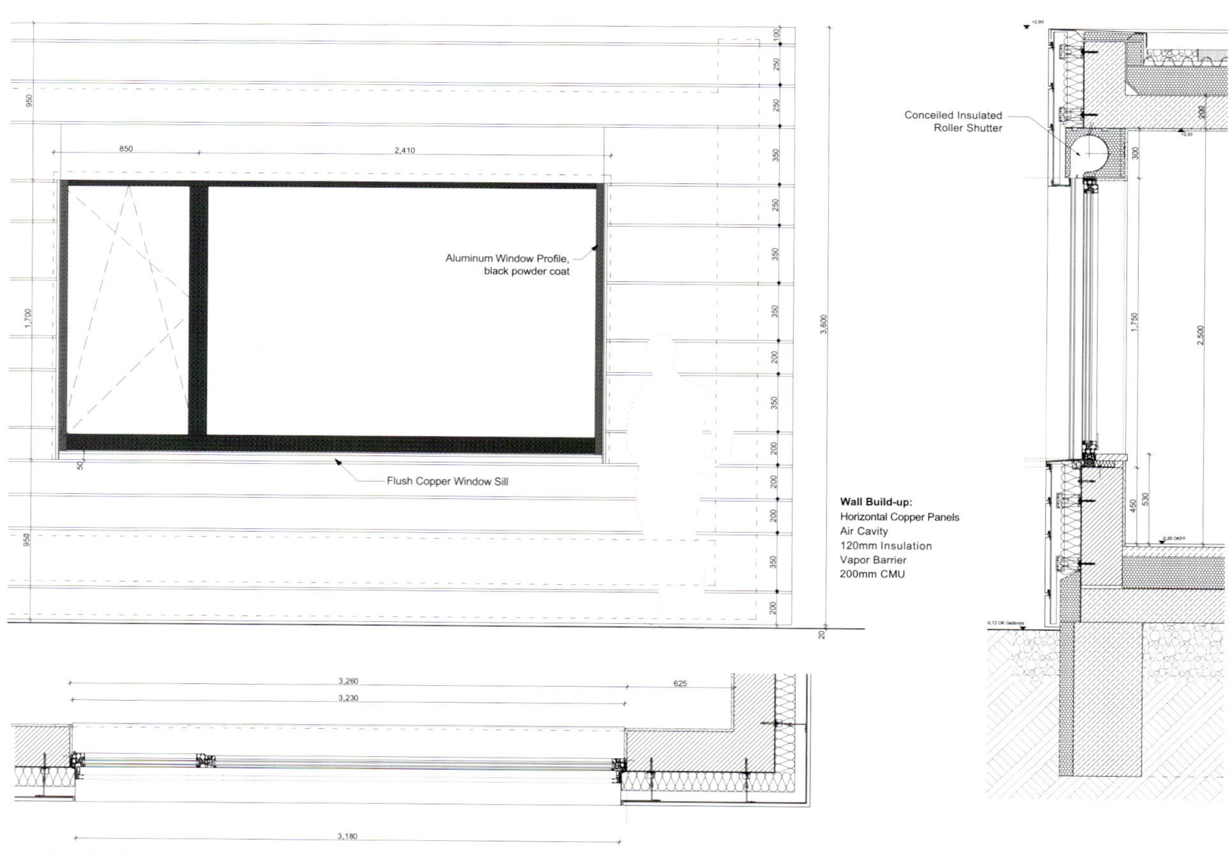

Copper Facade (plan, section, elevation)

© Anne Menke

© Anne Menke

HOUSE IN MUZZANO, Ticino, Switzerland

DAVIDE MACULLO ARCHITECTS

Idea Sketch

© Pino Musi

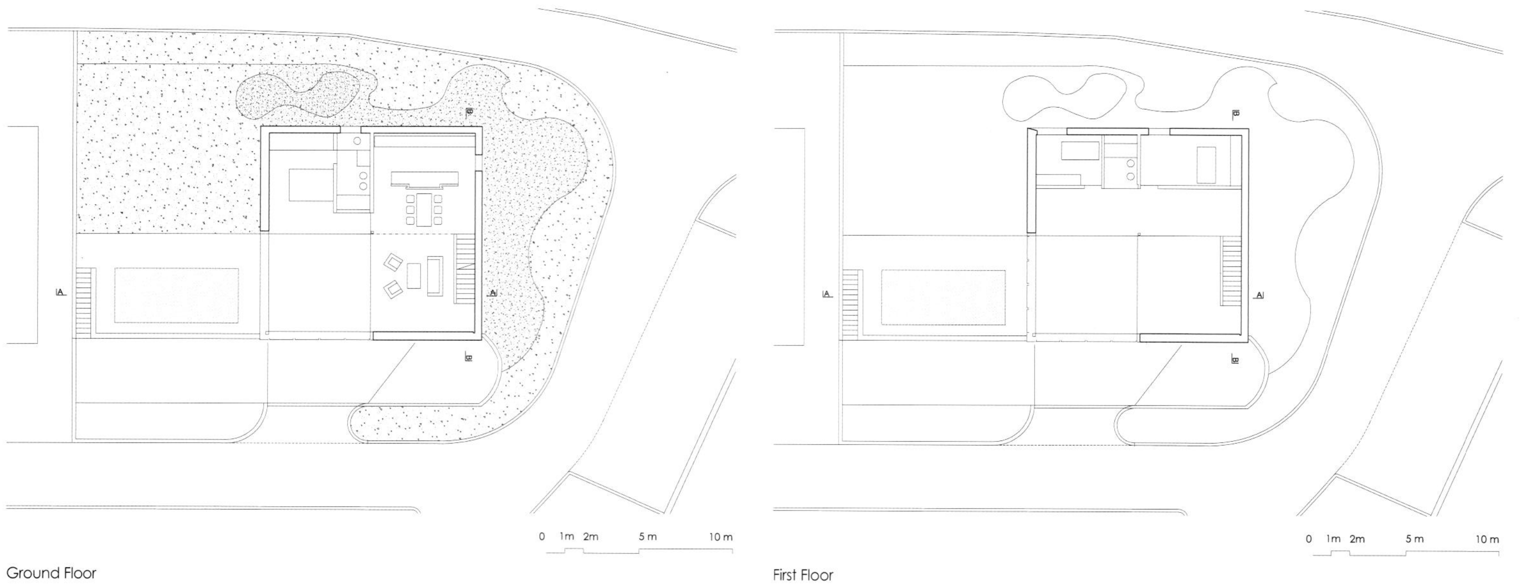

Ground Floor

First Floor

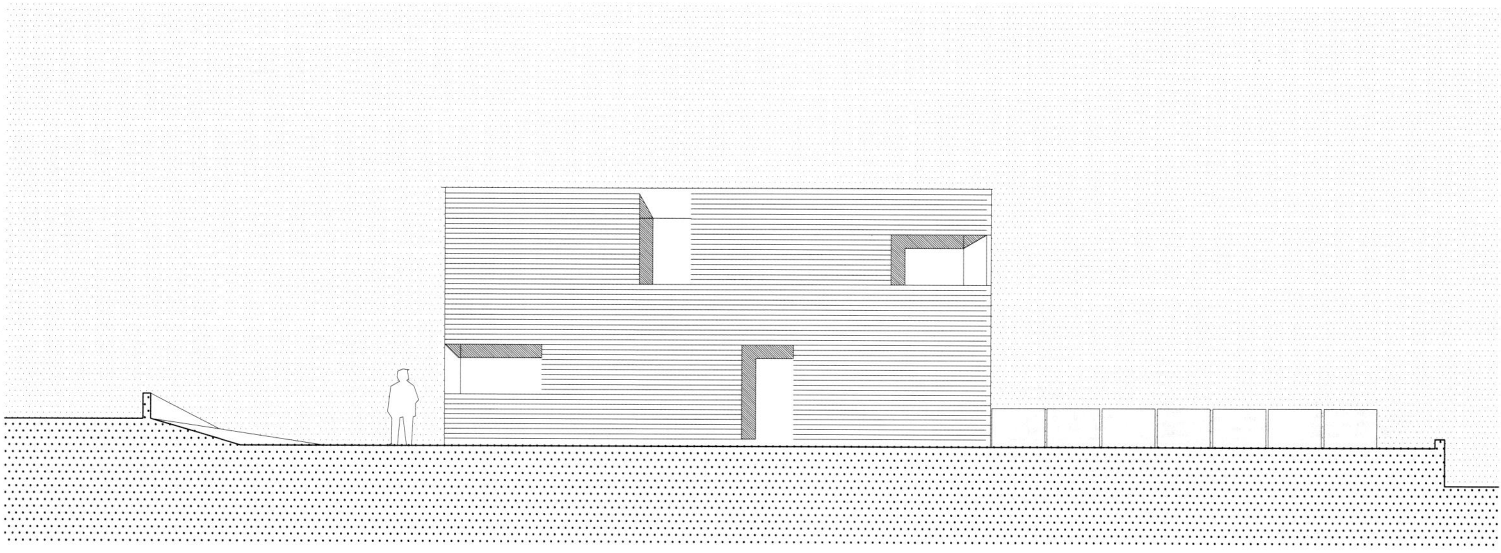

East Elevation

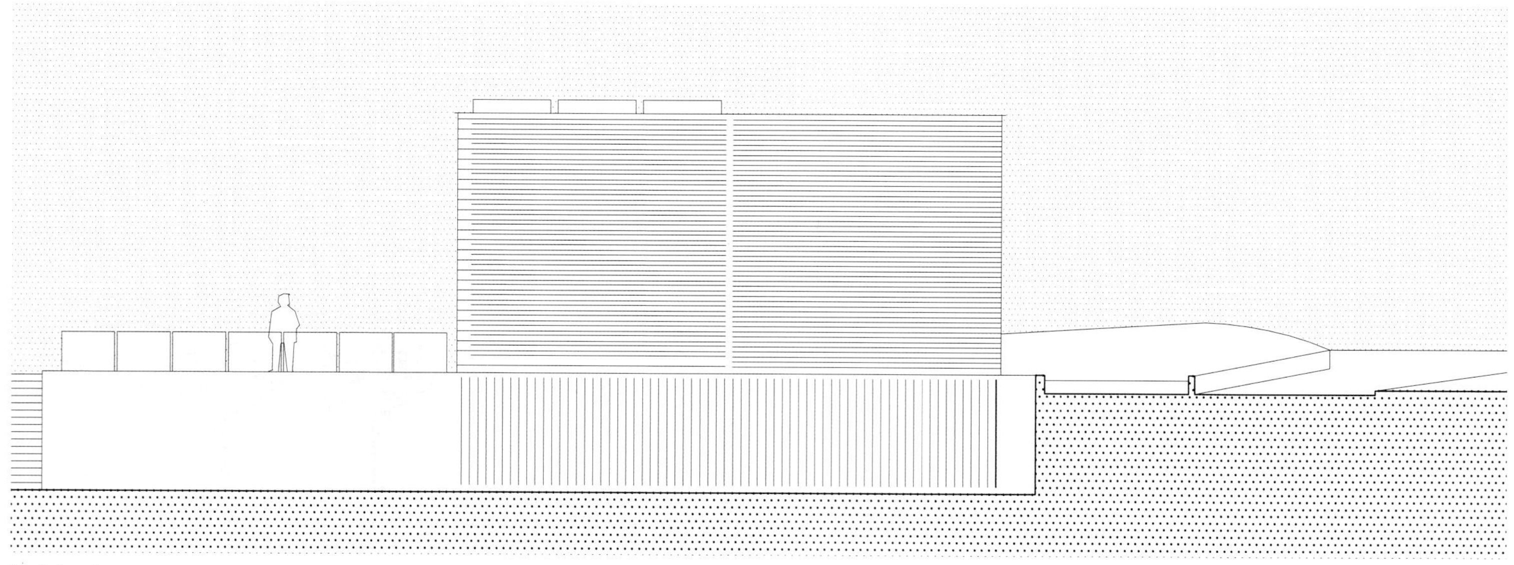

South Elevation

© Pino Musi

© Pino Musi

© Pino Musi

© Pino Musi

© Pino Musi

© Pino Musi

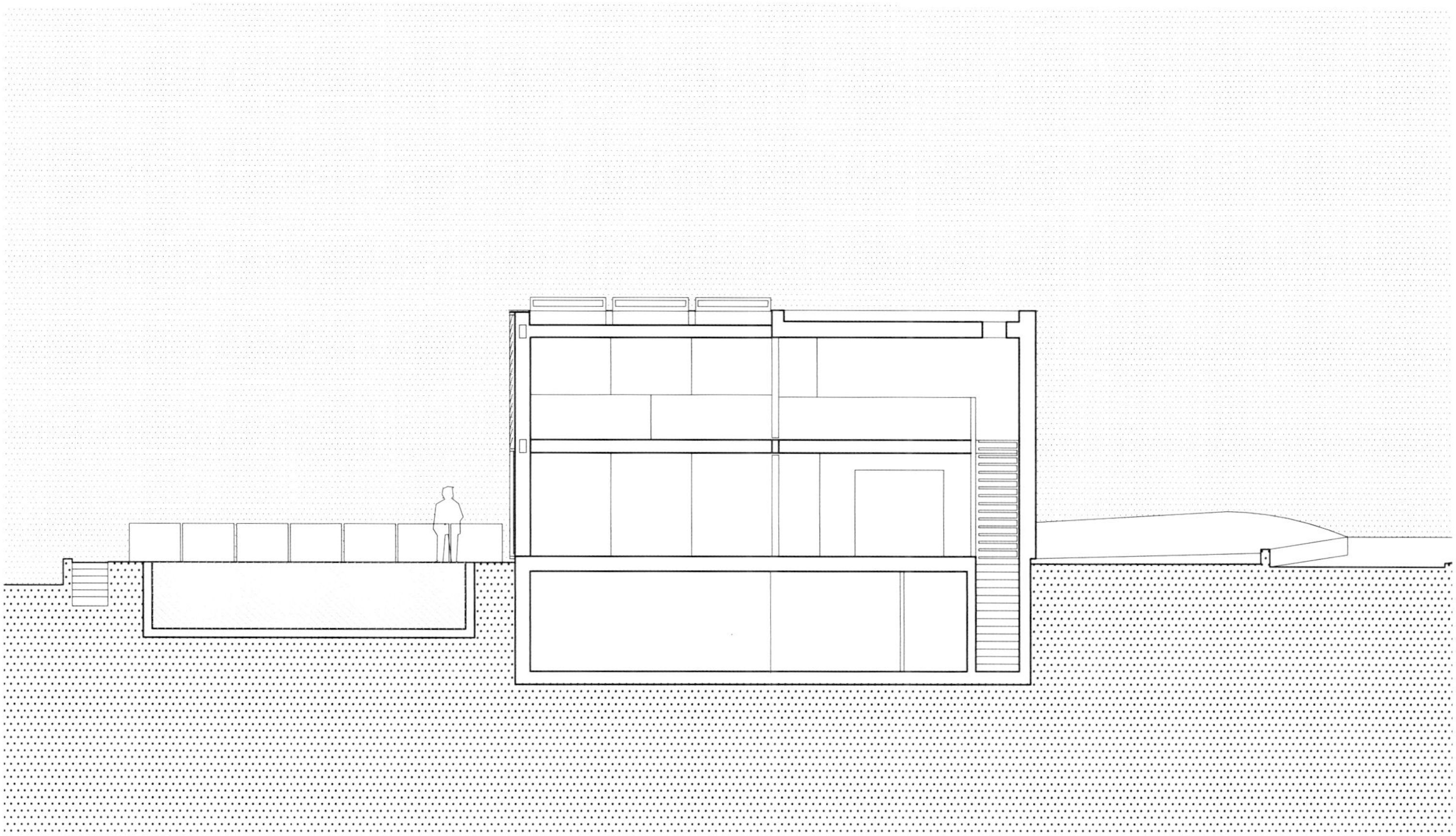

Section A-A

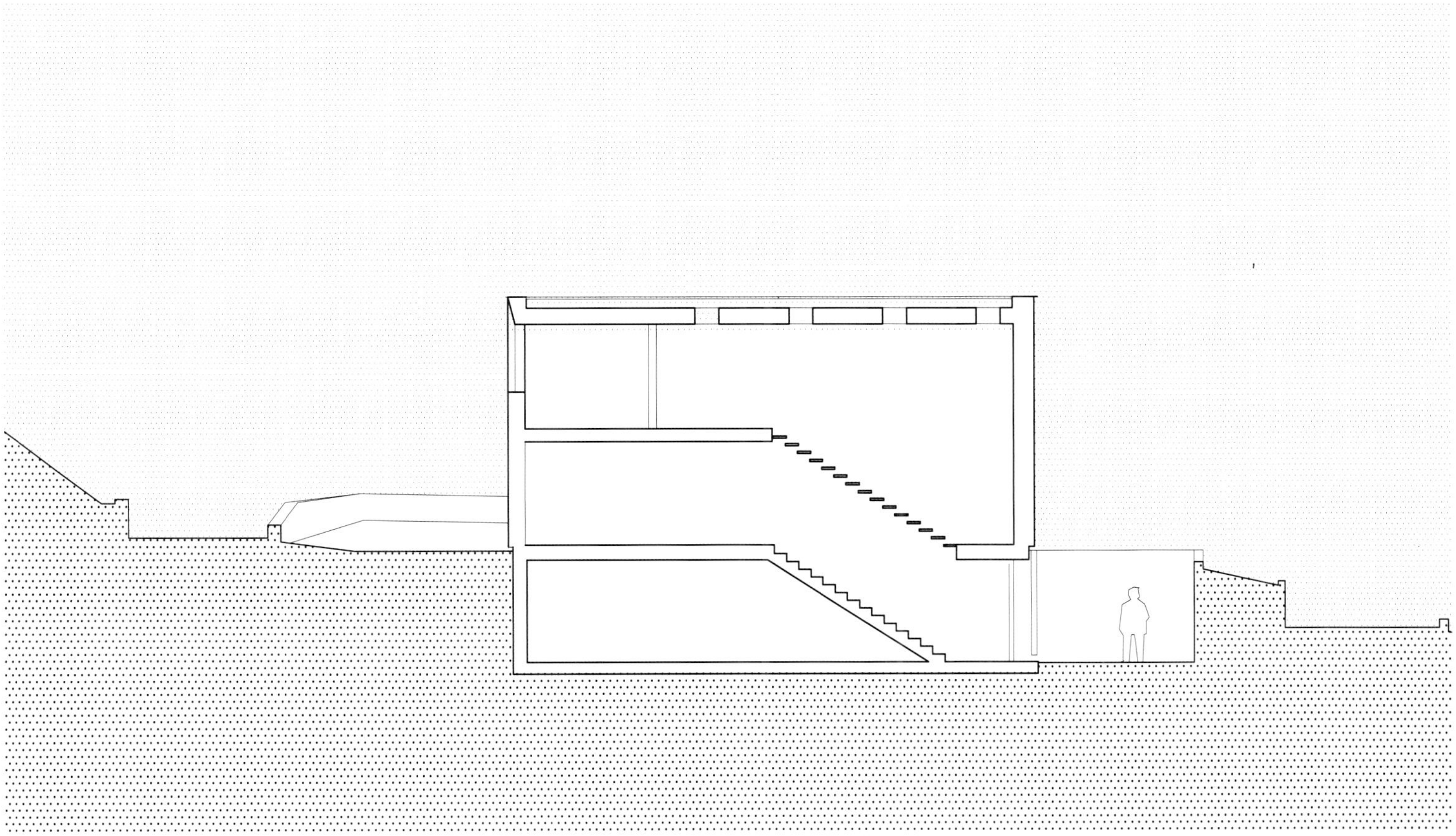

Section B-B

© Pino Musi

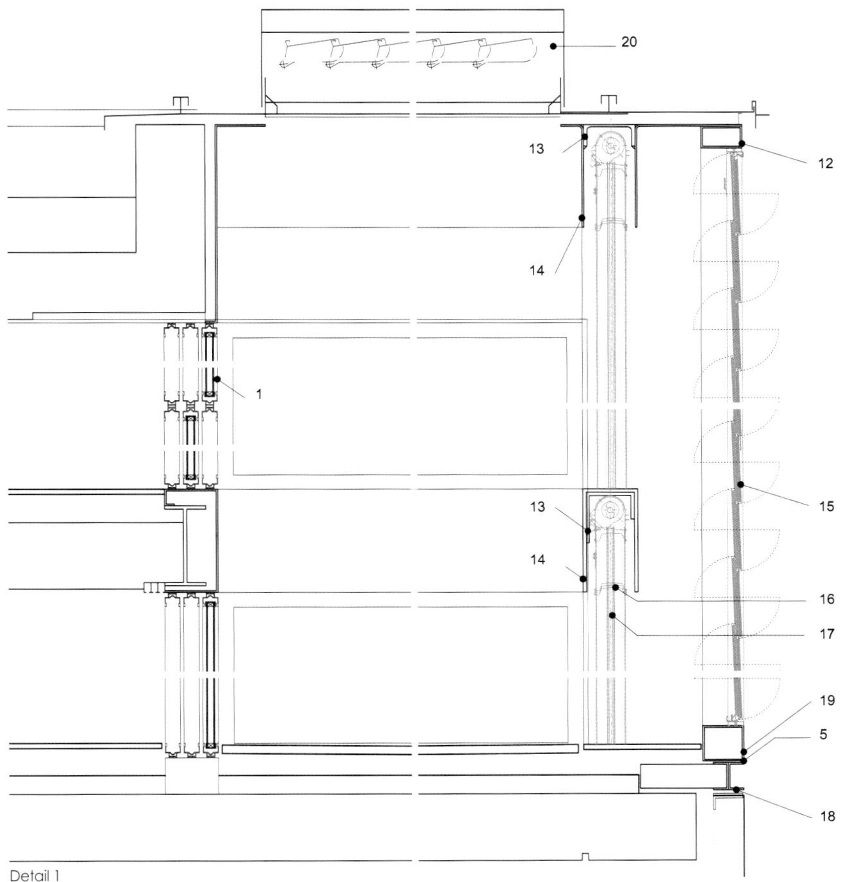

Detail 1

LEGEND

1 SLIDING GLASS WINDOWS TYPE
VITROCSA 3001 FRAME ALUMINIUM ANODIZED
5 ANTI OXIDATION LAYER
12 ALUMINIUM PROFILE ANODIZED
13 STRUCTURAL STEEL "U" PROFILE
14 STEEL COVER TOOL RAL 9006
15 ELECTRICAL GLASS LOUVERS TYPE
HAHN FRAME ALUMINIUM ANODIZED
16 ELECTRICAL ALUMINIUM LOUVERS
17 ALUMINIUM RAILS
18 "U" STRUCTURAL STEEL PROFILE 80/50 MM
19 PROFILE 120/105 MM ALUMINIUM ANODIZIED
20 SMOKE EXHAUST ELECTRICAL BOXES (LOUVERS)
TYPE "COLT"

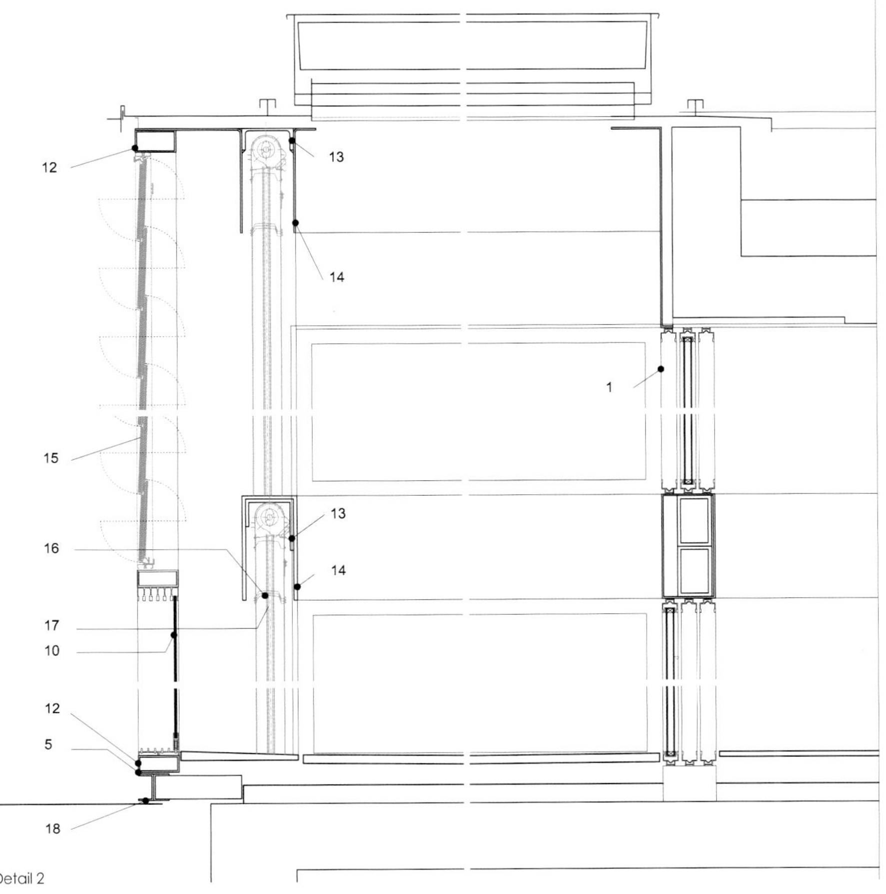

Detail 2

LEGEND

1 SLIDING GLASS WINDOWS TYPE
VITROCSA 3001 FRAME ALUMINIUM ANODIZED
5 ANTI OXIDATION LAYER
10 GLASS SLIDING PANELS TYPE
AWESO 330 FRAME ALUMINIUM ANODIZIED
12 ALUMINIUM PROFILE ANODIZED
13 STRUCTURAL STEEL "U" PROFILE
14 STEEL COVER TOOL RAL 9006
15 ELECTRICAL GLASS LOUVERS TYPE
HAHN FRAME ALUMINIUM ANODIZED
16 ELECTRICAL ALUMINIUM LOUVERS
17 ALUMINIUM RAILS
18 "U" STRUCTURAL STEEL PROFILE 80/50 MM

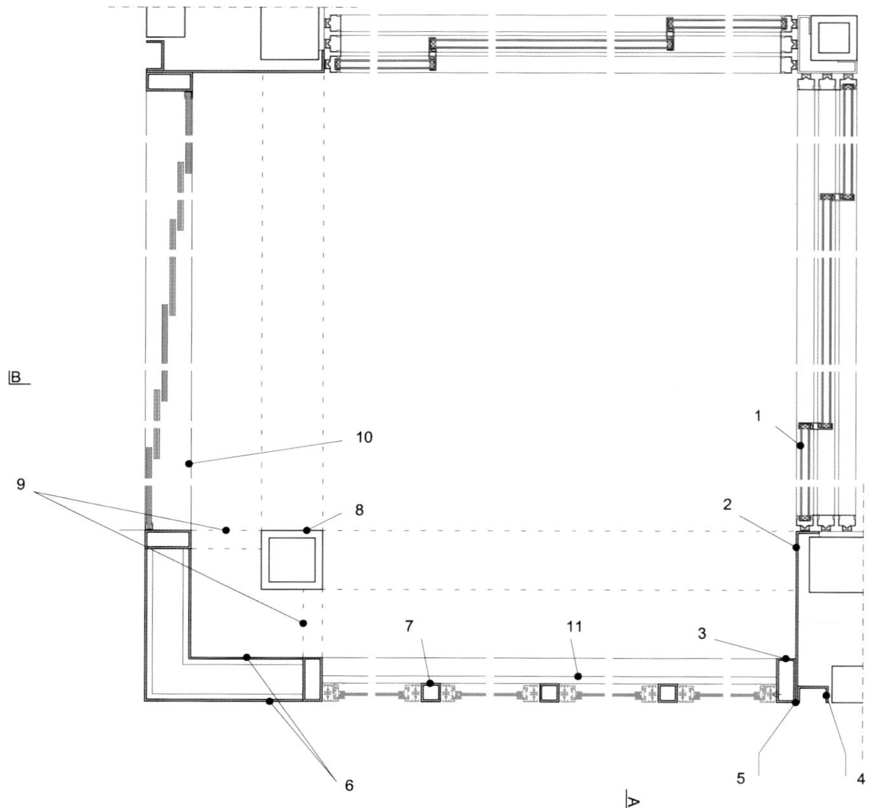

LEGEND

1 SLIDING GLASS WINDOWS TYPE
VITROCSA 3001 FRAME ALUMINIUM ANODIZED
2 LOVER STEEL TOOL RAL 9006
3 PROFILE 50/120 MM ALUMINIUM ANODIZED
4 STEEL PROFILE "U" 80/50 MM RAL 9006
5 ANTI OXIDATION LAYER
6
7 PROFILE 50/50 MM ALUMINIUM ANODIZIED
8 STRUCTURE 150/150 MM RAL 9006
9 FIXATION STEEL 20/20 MM RAL 9006
10 GLASS SLIDING PANELS TYPE
AWESO 330 FRAME ALUMINIUM ANODIZIED
11 STEEL PROFILE 20 MM RAL 9006

Detail 3

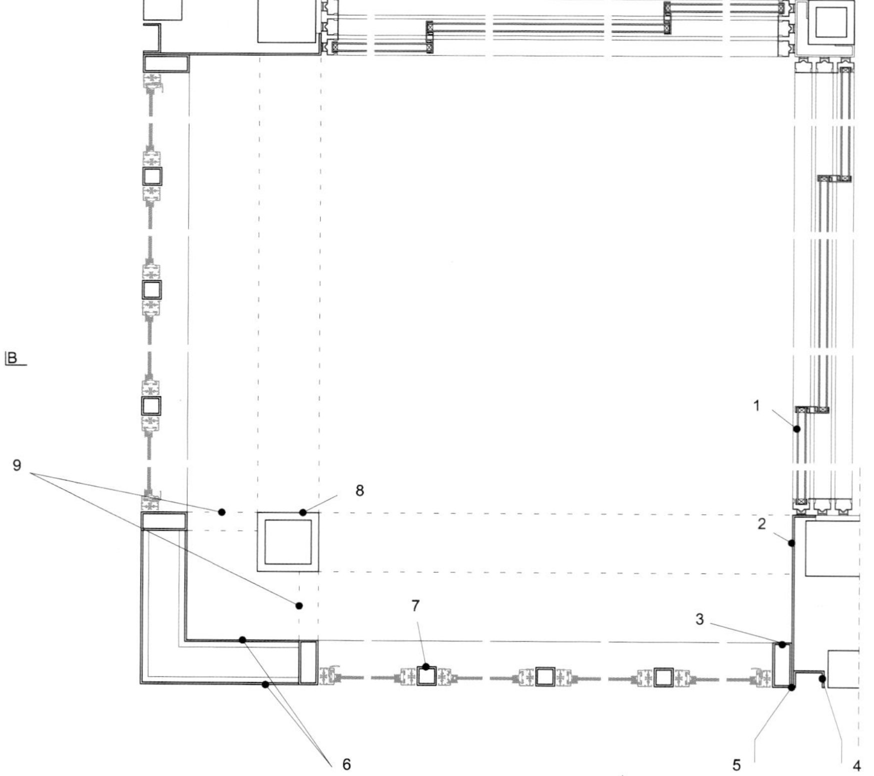

LEGEND

1 SLIDING GLASS WINDOWS TYPE
VITROCSA 3001 FRAME ALUMINIUM ANODIZED
2 LOVER STEEL TOOL RAL 9006
3 PROFILE 50/120 MM ALUMINIUM ANODIZED
4 STEEL PROFILE "U" 80/50 MM RAL 9006
5 ANTI OXIDATION LAYER
6 ALUMINIUM TOOL ANODIZIED
7 PROFILE 50/50 MM ALUMINIUM ANODIZIED
8 STRUCTURE 150/150 MM RAL 9006
9 FIXATION STEEL 20/20 MM RAL 9006

Detail 44

© Pino Musi

Love House, Madrid, Spain

nodo17 Architects

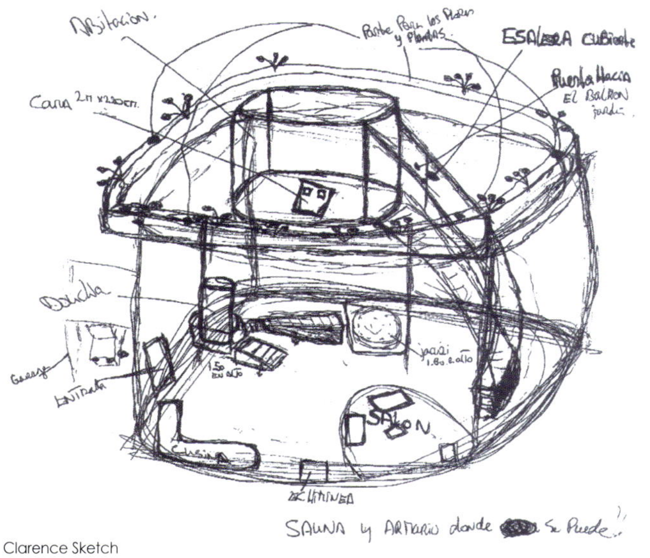

Clarence Sketch

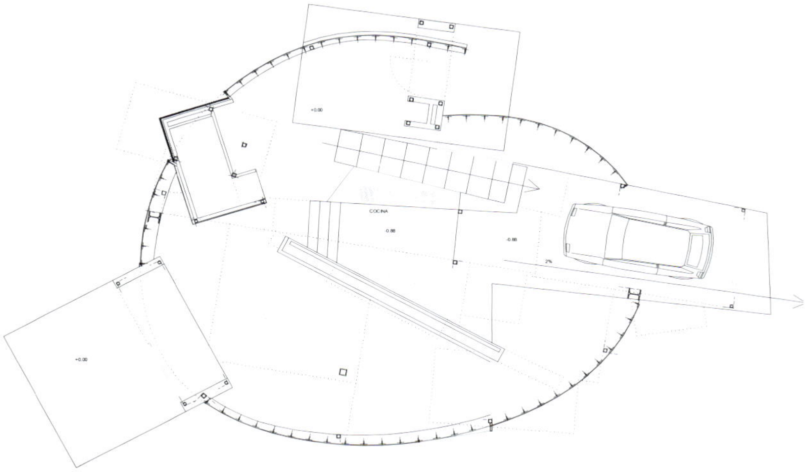

Ground Floor Level

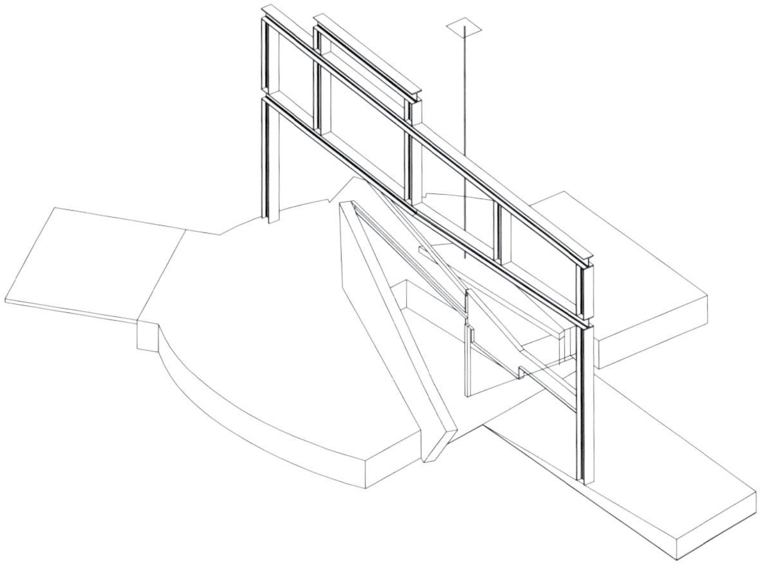

Axonometric

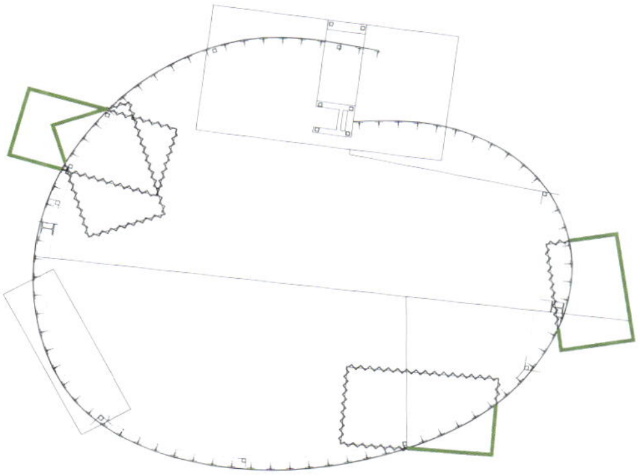

Hairy Boxes - Exterior: Vegetation Panels/ Interior: Hairy Carpet

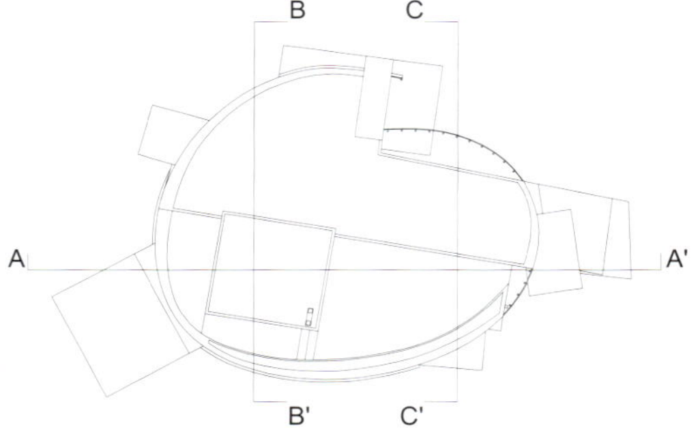

Spread screen elevation

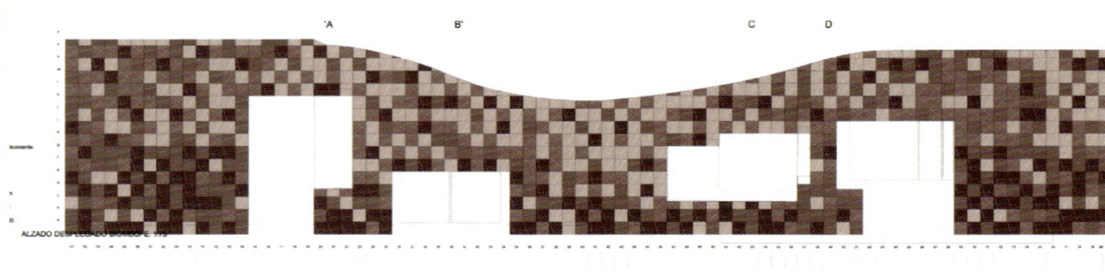

Bark

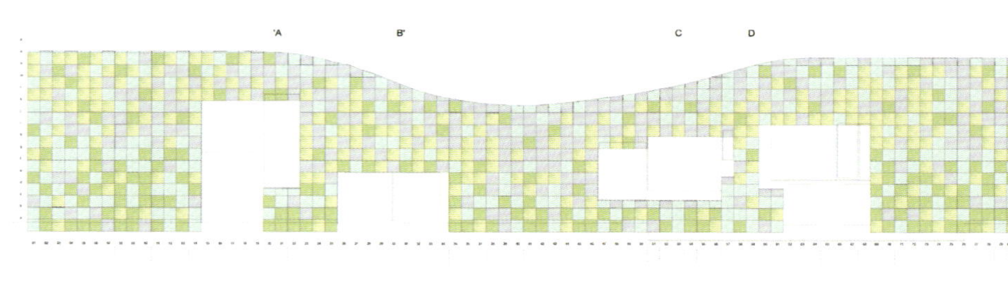

Camoufagle

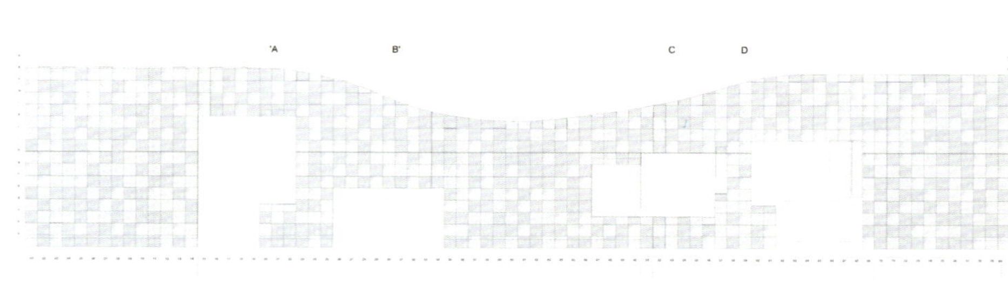

Cloud

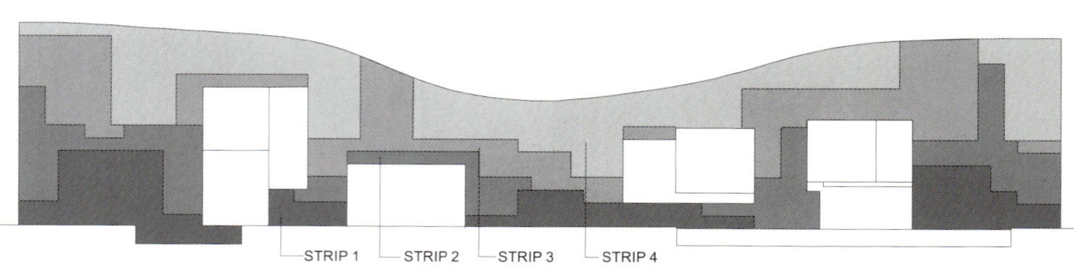

DENSITY BANDS/ Facade divisions

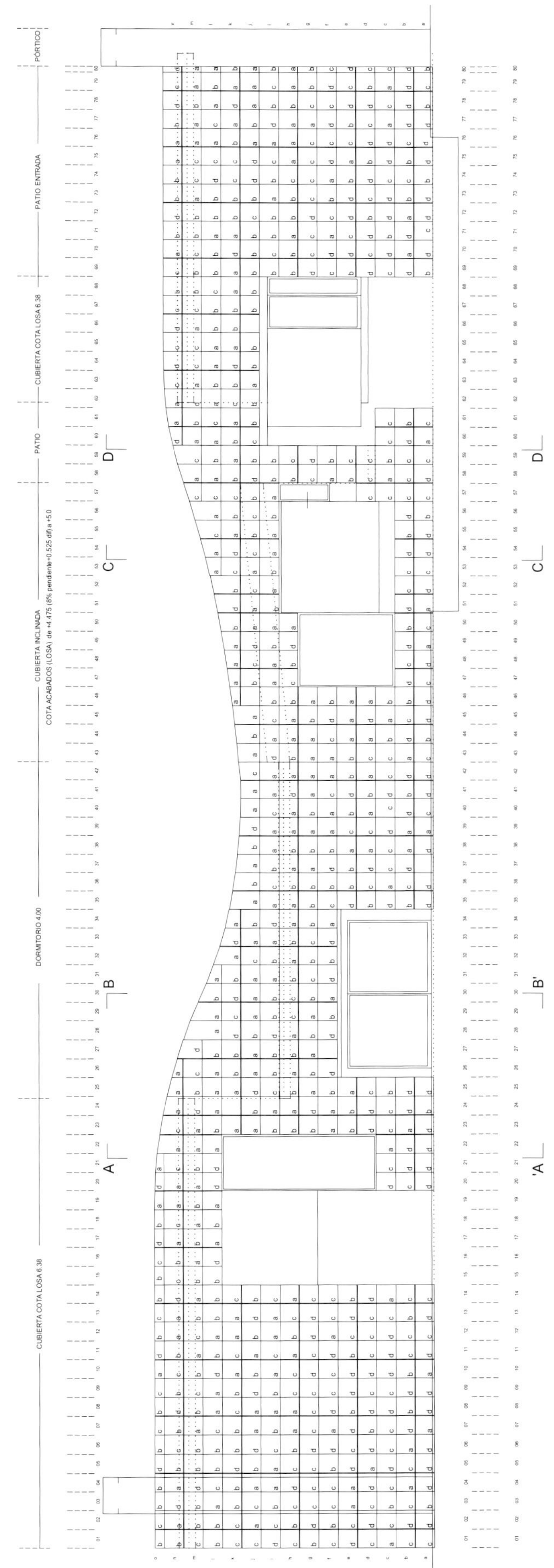
PORTICO
PATIO ENTRADA
CUBIERTA COTA LOSA 6.38
PATIO
CUBIERTA INCLINADA
DORMITORIO 4.00
CUBIERTA COTA LOSA 6.38

Maison Leguay, France
Jacques Moussafir

© Jérôme Ricolleau

Section of the Extension

1. black anthra-zinc cover
2. wooden boarding
3. wooden skeleton
4. black anthra-zinc gutter
5. glasswool insulation - 200mm
6. plasterboard
7. black bricks - 50x110x210
8. larch glulam window frame
9. raw concrete frame
10. foam insulation
11. raw concrete breeze blocks
12. steel support for the bricks
13. glasswool insulation - 120mm
14. concrete slab
15. plasterboard
16. black anthra-zinc cover
17. larchwood sliding door
18. steel column
19. heating floor
20. concrete floor covering
21. watertightness

© Hervé Abbadie

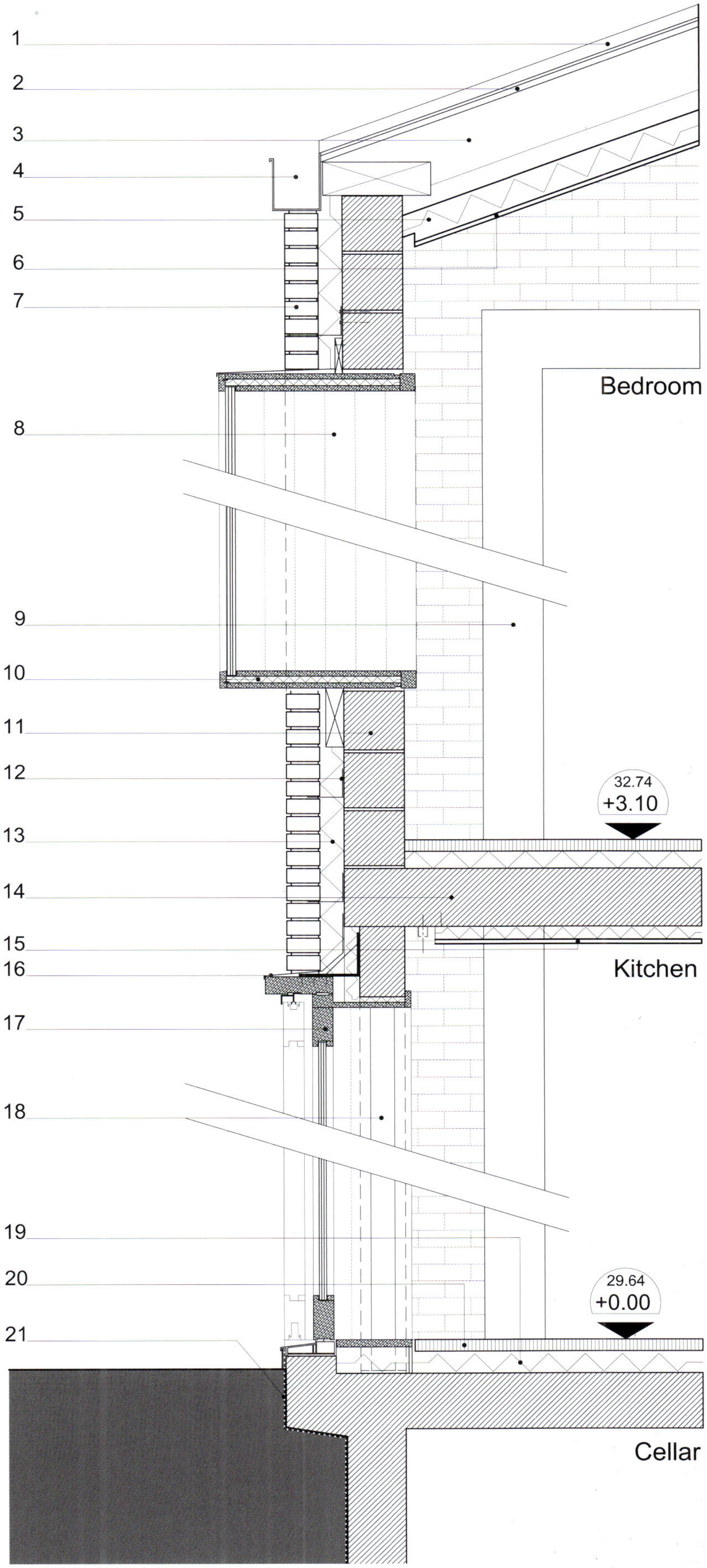

1
2
3
4
5
6
7
8
9
10
11
12
13
14
15
16
17
18
19
20
21
Bedroom
32.74
+3.10
Kitchen
29.64
+0.00
Cellar

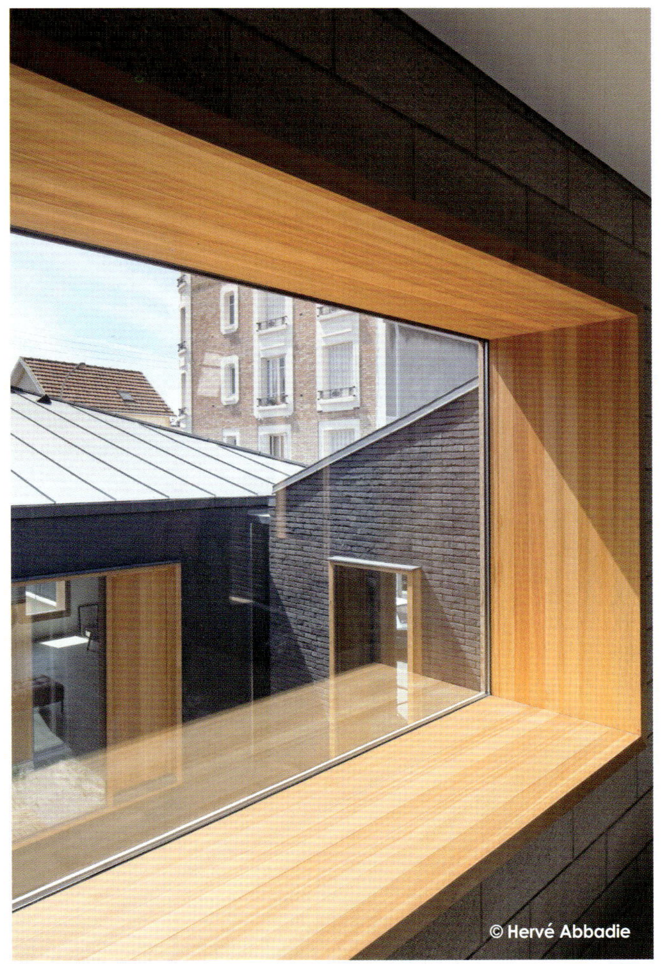

© Jérôme Ricolleau

© Jérôme Ricolleau

© Jérôme Ricolleau

MJ, Hyogo, Japan
Keiichi Hayashi Architect

© Yoshiyuki Hirai

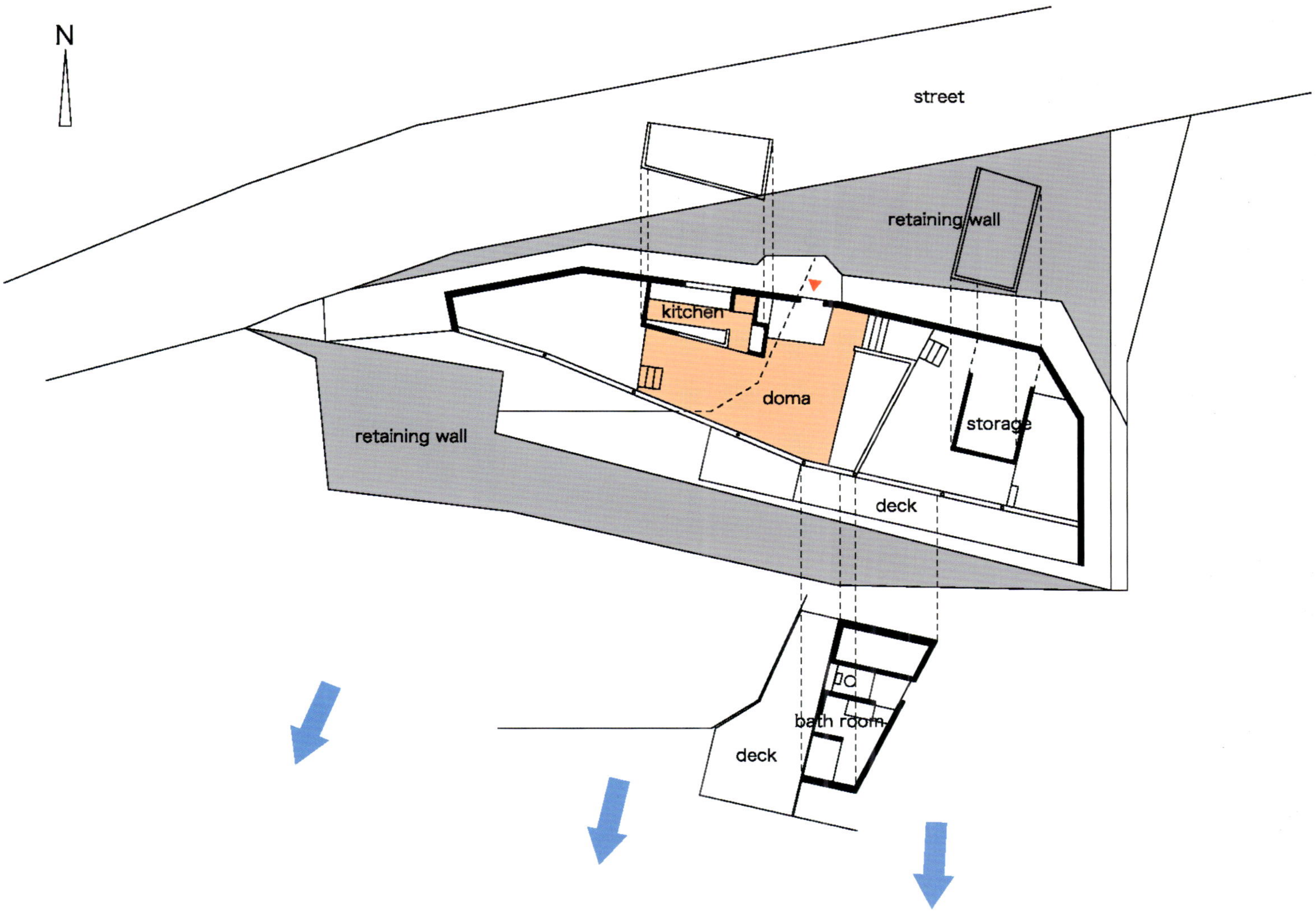

N
street
retaining wall
kitchen
doma
storage
retaining wall
deck
bath room
deck
seto inland sea

Site Plan

Seto inland Sea
Sandy beach
JR Sanyo Line
Route 2 of national highway
Sanyo railway
Houses
MJ
Houses
Houses
Frontal road

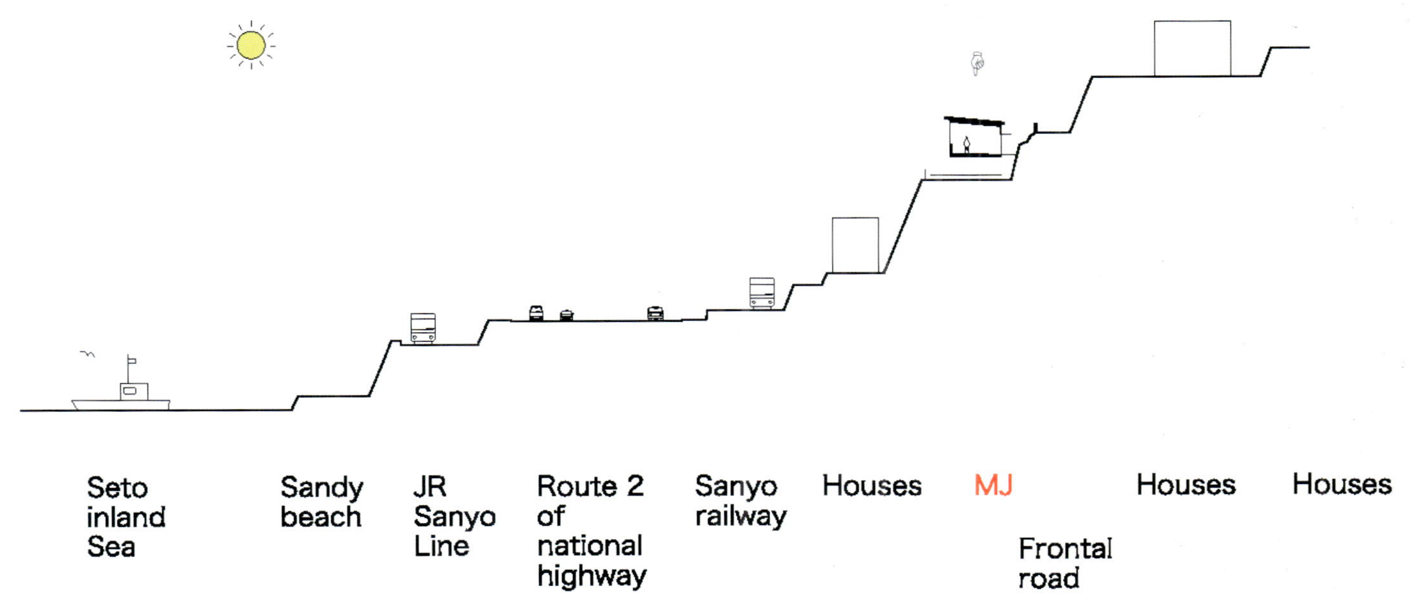

Site Section

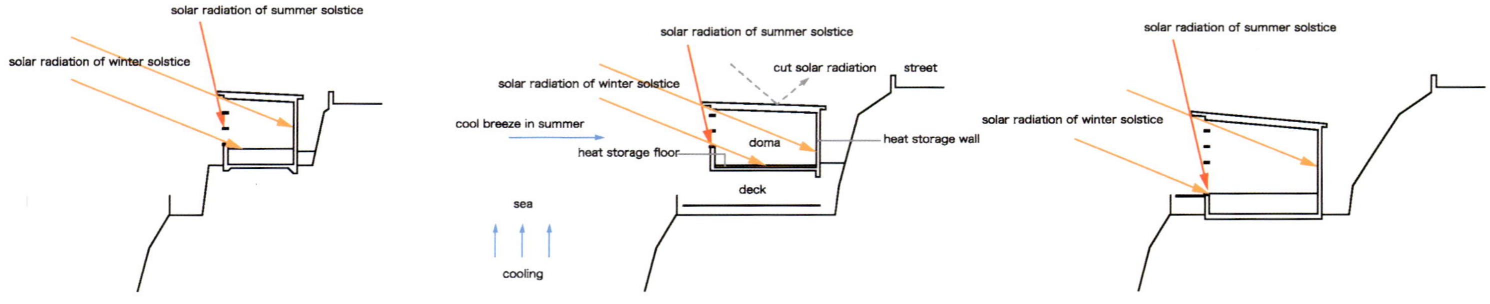

Section A / B / C

Construction Process

© Yoshiyuki Hirai

© Yoshiyuki Hirai

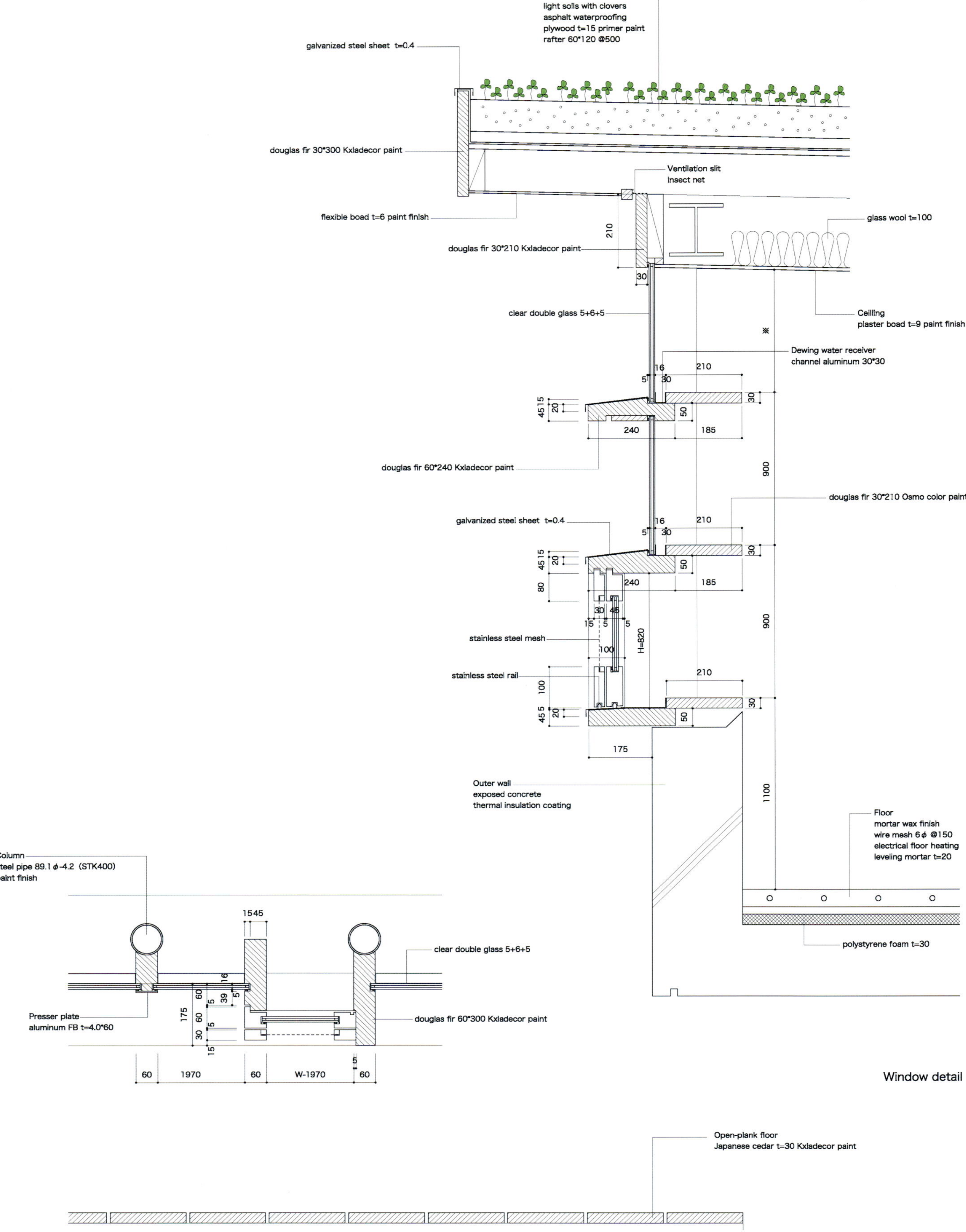

Construction Detail

Mountain Retreat, Bolzano, Italy

modostudio

© Laura Egger

© Laura Egger

© Laura Egger

© Laura Egger

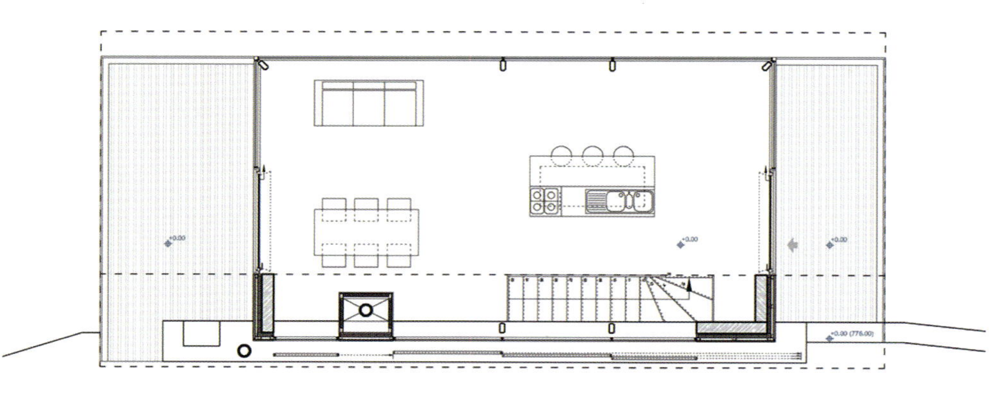

First Floor

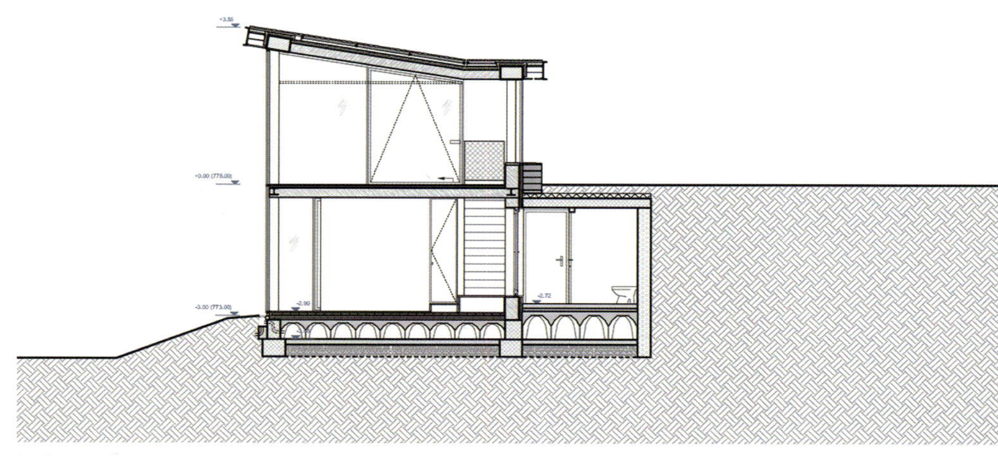

Section

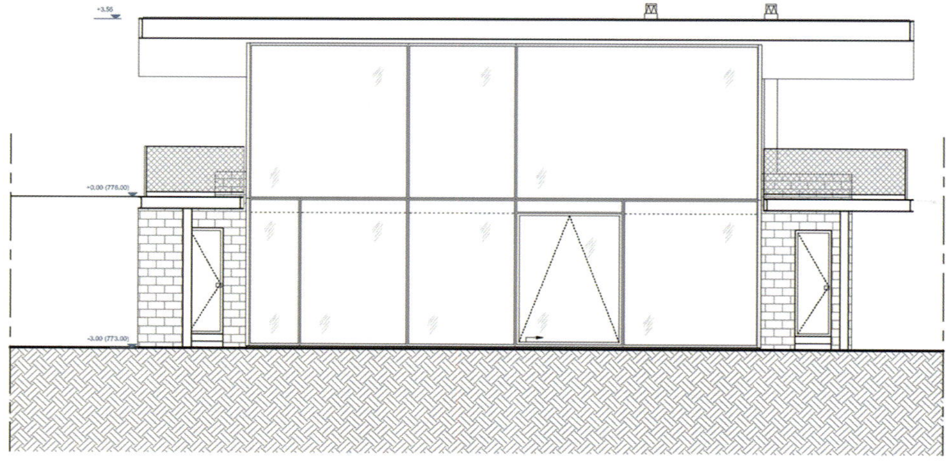

Elevation

© Laura Egger

© Laura Egger

MR House, Pomponne, France

PERIPHERIQUES

Color Concept

Facade

Door Detail

Muslhaufen House, Luson, Italy
Armin Blasbichler Studio

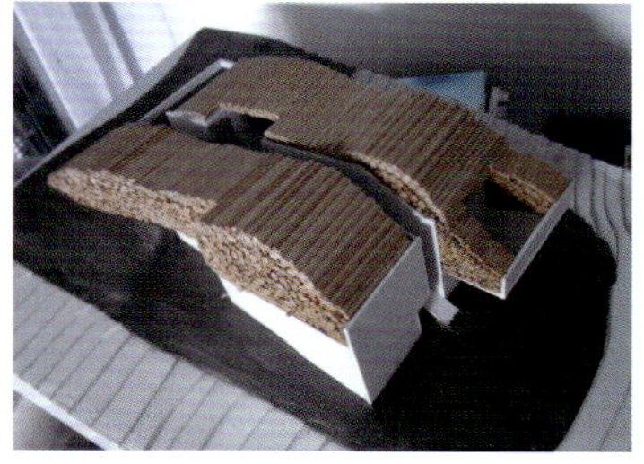

Study Modeling

V-House, Oslo, Norway

SPACEGROUP

© Jeroen Musch & Ivan Brodey

© Jeroen Musch & Ivan Brodey

© Jeroen Musch & Ivan Brodey

© Jeroen Musch & Ivan Brodey

Apartment House Gradaska, Ljubljana, Slovenia

Sadar Vuga Arhitekti

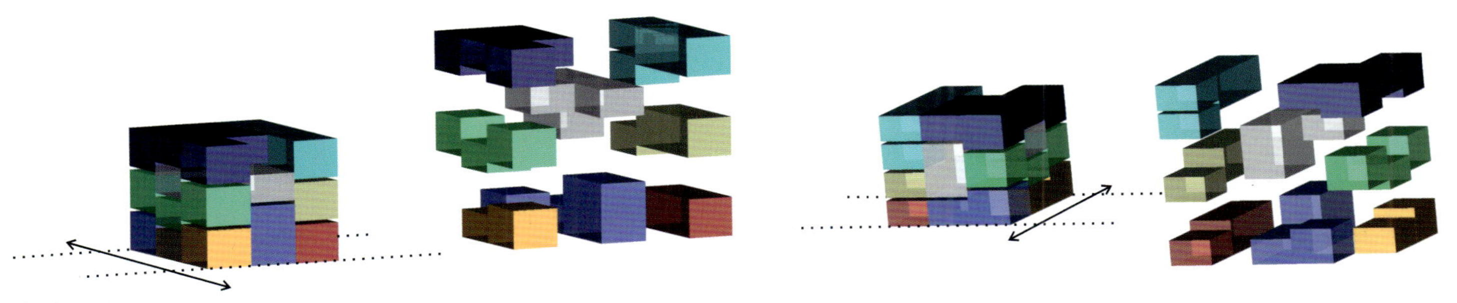

Idea Concept

Elevation

Baroque Court Apartments, ljubljana, slovenia

OFIS arhitekti

©Tomaz Gregoric, Jan Celeda

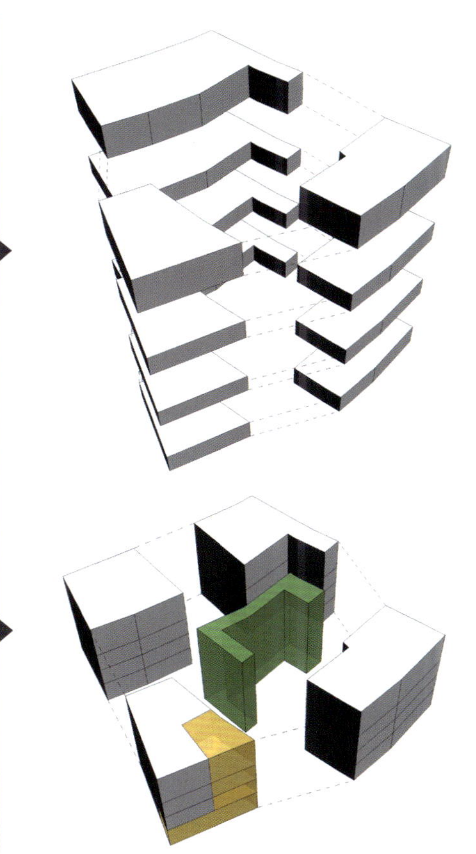

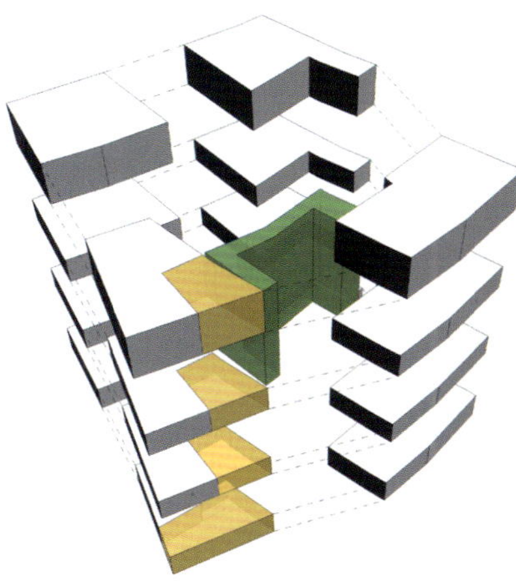

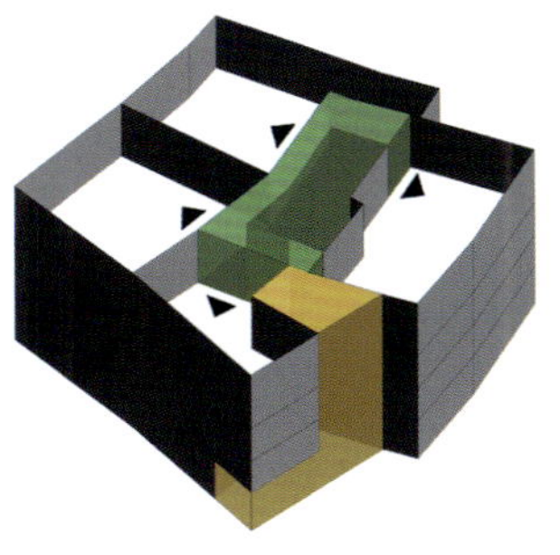

connecting court

communications

Old Street

Before → After

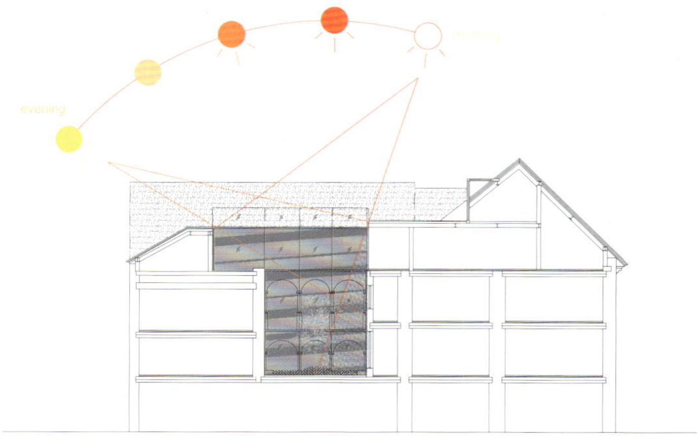

Section B-B

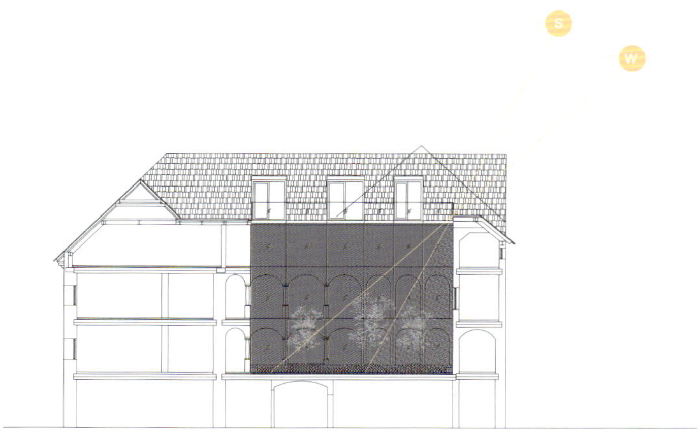

Section C-C

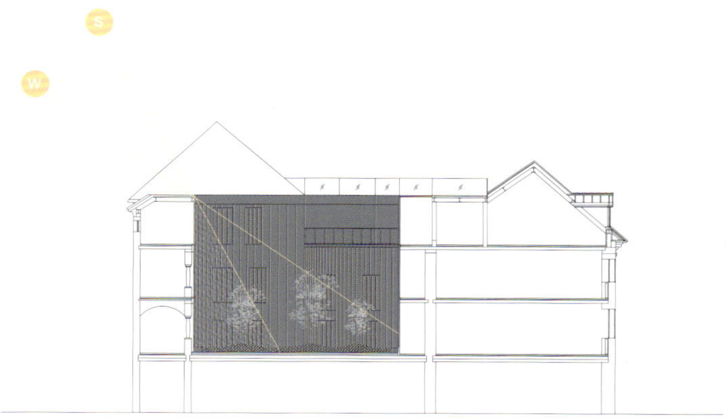

Section A-A

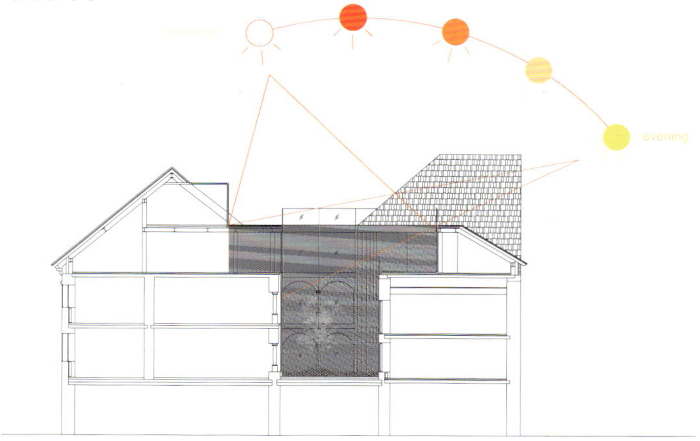

Section D-D

© Tomaz Gregoric, Jan Celeda

© Tomaz Gregoric, Jan Celeda

© Tomaz Gregoric, Jan Celeda

© Tomaz Gregoric, Jan Celeda

© Tomaz Gregoric, Jan Celeda

© Tomaz Gregoric, Jan Celeda

© Tomaz Gregoric, Jan Celeda

© Tomaz Gregoric, Jan Celeda

CO2 Neutral Housing Blocks, Antwerp, Belgium

Casanova+Hernandez Architects

© CHA

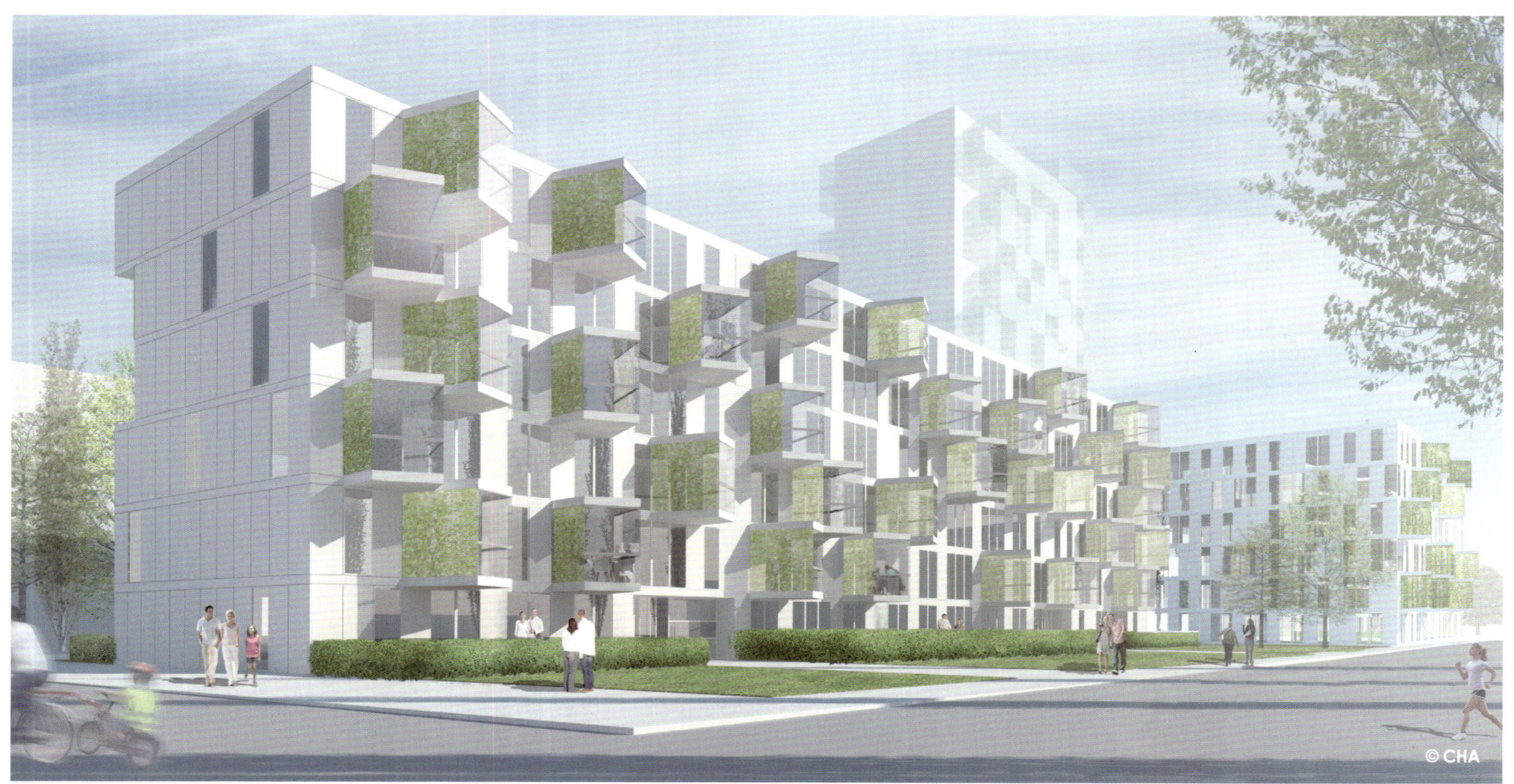

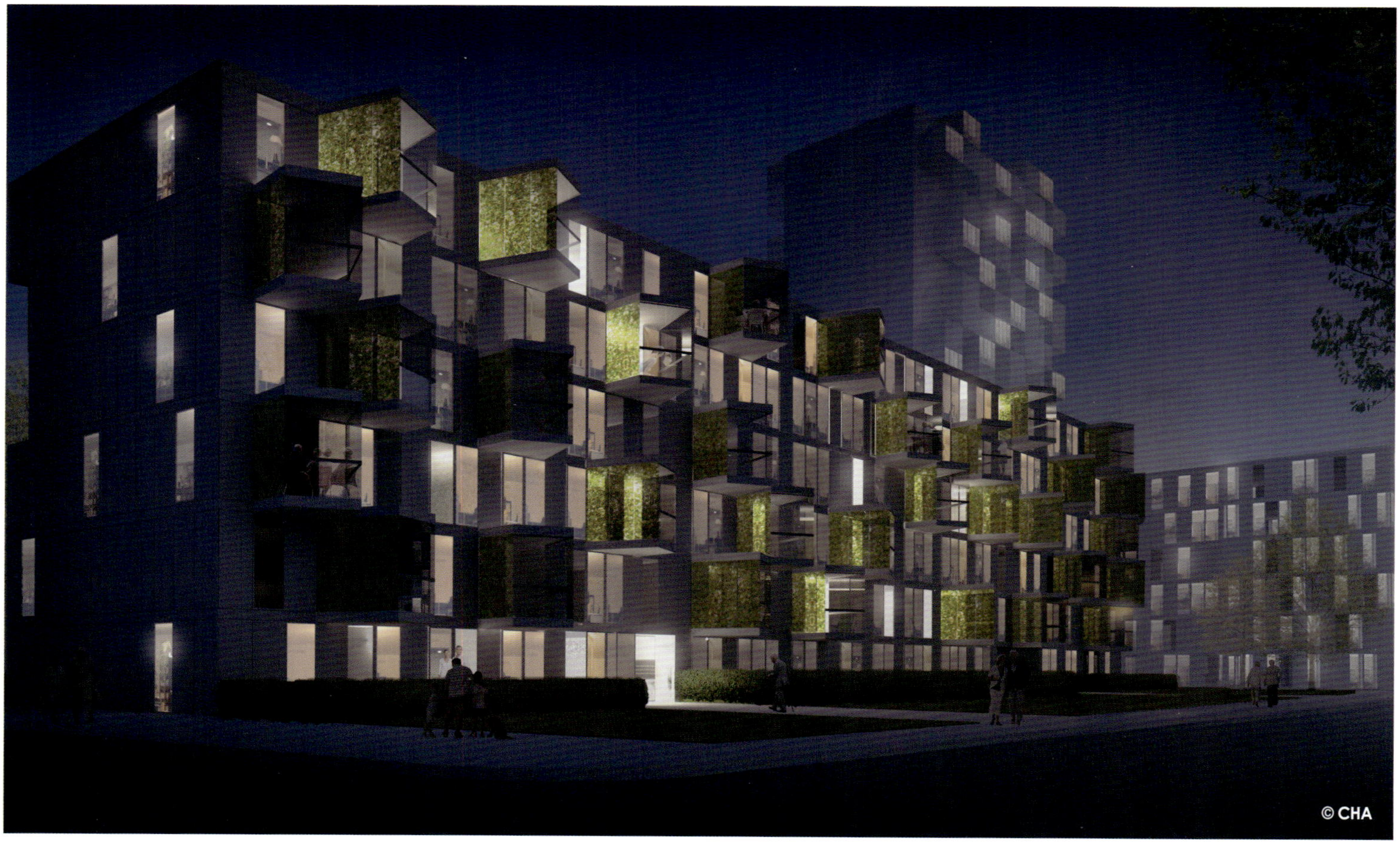

© CHA

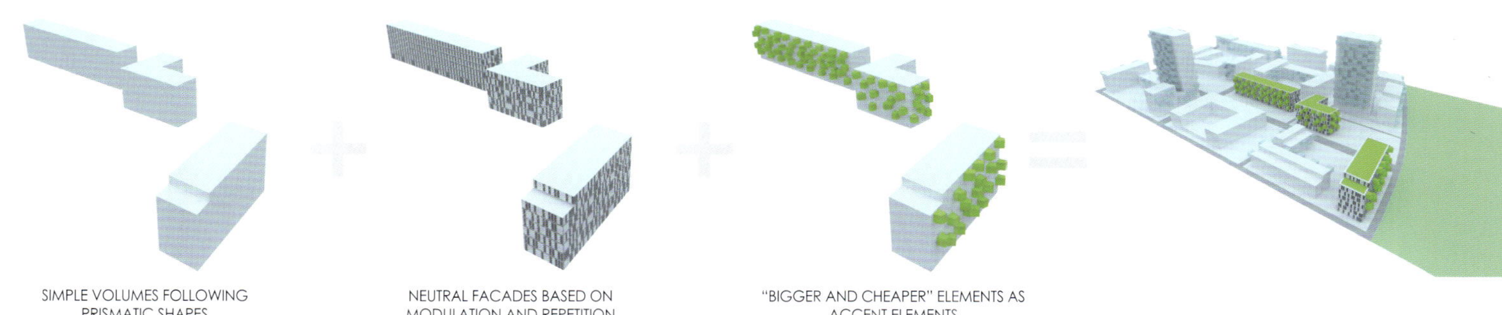

SIMPLE VOLUMES FOLLOWING PRISMATIC SHAPES

NEUTRAL FACADES BASED ON MODULATION AND REPETITION

"BIGGER AND CHEAPER" ELEMENTS AS ACCENT ELEMENTS

Volume and Skin Study

SUNLIGHT ORIENTED

VIRTUAL FOREST AS BUILDING EXPRESSION

IDENTITY OF EACH BUILDING RELATED TO AN AUTOCHTHONOUS TREE

Site Analysis

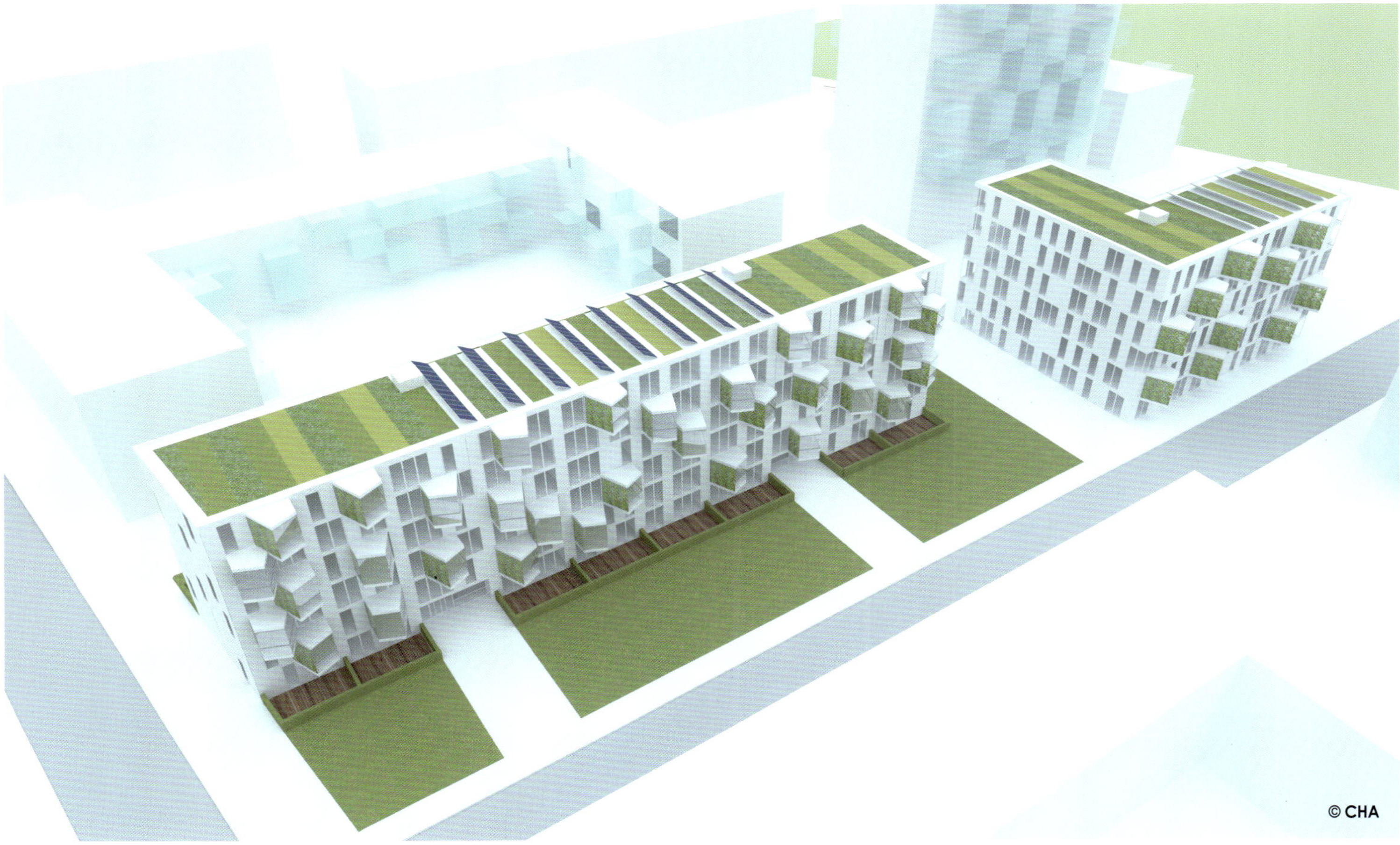

© CHA

CONSTRUCTIVE ANALYSIS

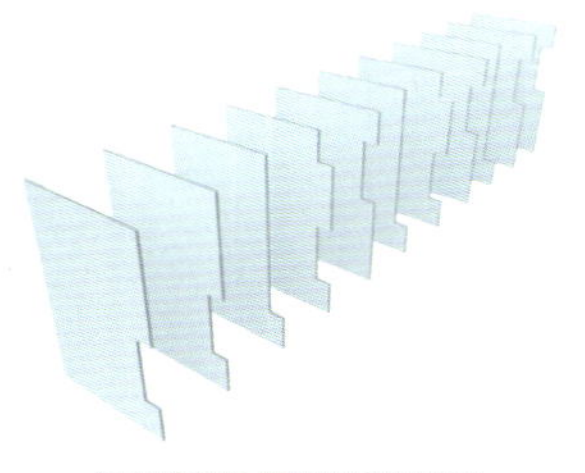

CONCRETE BEARING WALLS

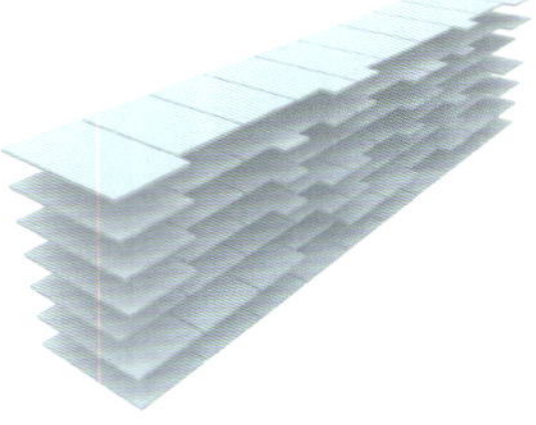

CONCRETE PREFABRICATED SLABS

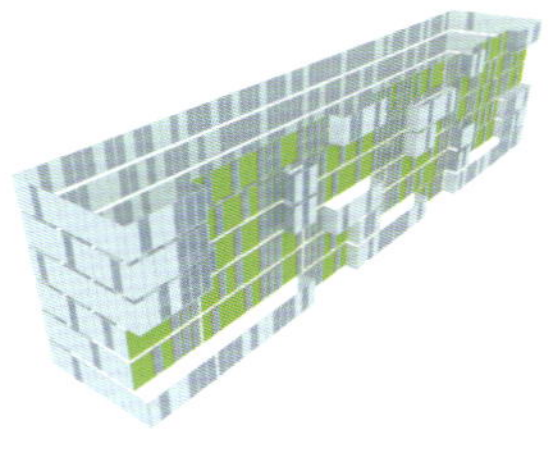

MODULATED ALUMINIUM FACADE, PANELS AND FRAMES

PREFABRICATED "BIGGER AND CHEAPER"

BIGGER AND CHEAPER

Parameters: Opening and closing of the glazed elements_Position and motive of the printed glass_Transparency level of printed glass

CLOSED

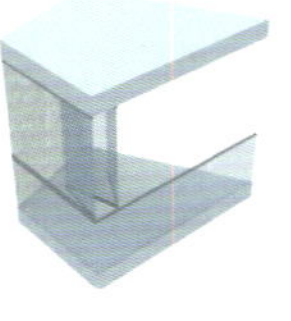

SEMI-OPEN

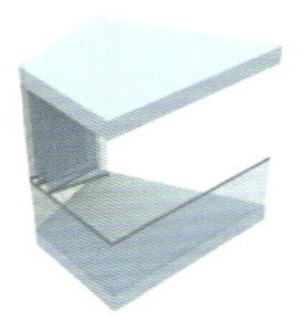

OPEN

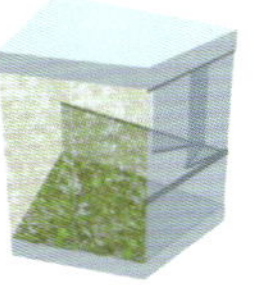

PRINTED GLASS

EXTERIOR APPEARANCE STUDY OF "BIGGER AND CHEAPER" FROM THE NORTH

Parameters: Number, shape and position of "Bigger and Cheaper" units_Material_Colour

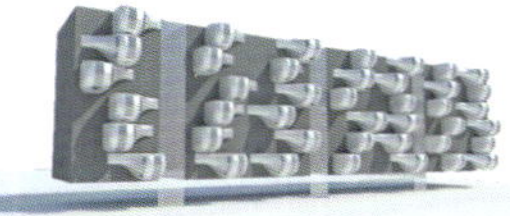
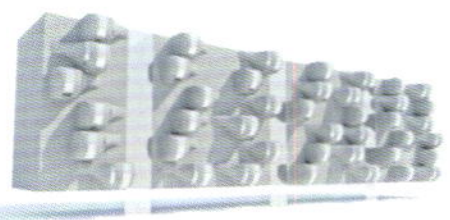

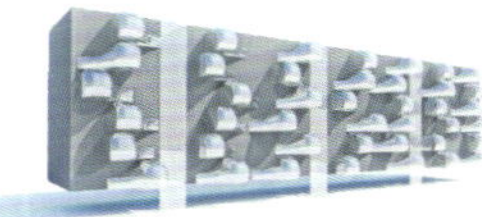

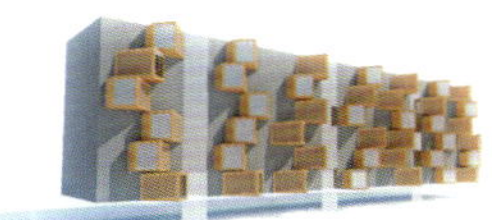

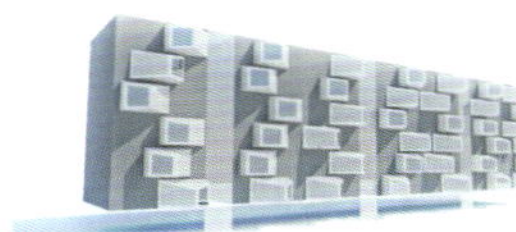

EXTERIOR APPEARANCE STUDY OF "BIGGER AND CHEAPER " FROM THE SOUTH

Parameters: Number, shape and position of "Bigger and Cheaper" units_Material_Colour_Printed glass pattern at south side

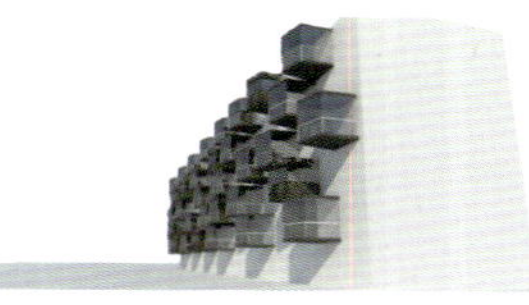

INTERIOR APPEARANCE AND LIGHT STUDY OF "BIGGER AND CHEAPER"

Parameters: Transparency_Visual relationship with the outside_Shadows_Pattern

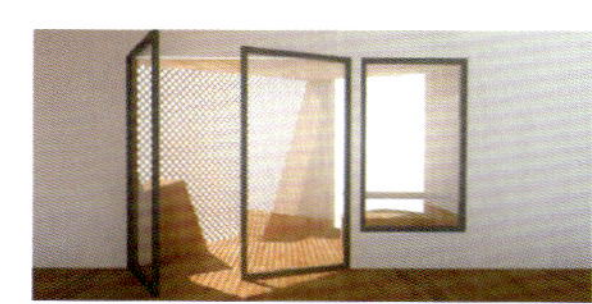

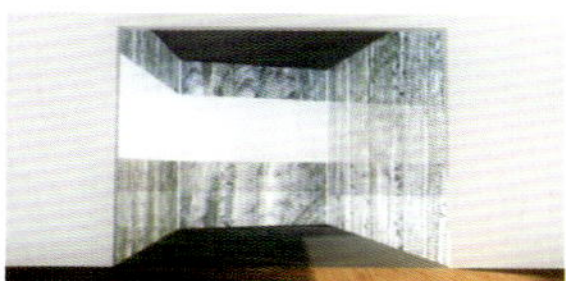

© CHA

© CHA

BEEKBERGEN, Blaricum, The Netherlands

Casanova+Hernandez Architects

© CHA

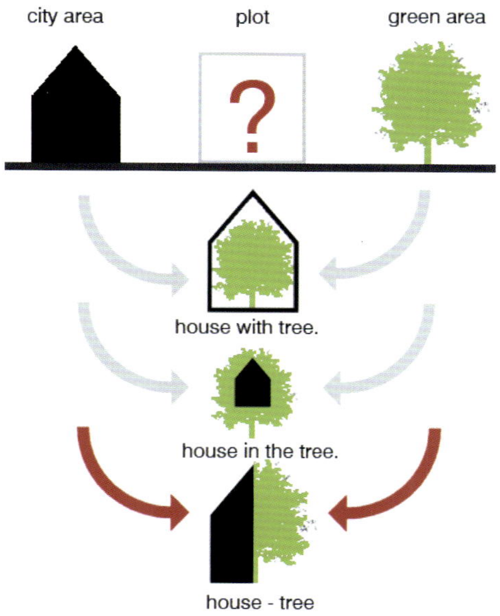

city area
plot
green area
?
house with tree.
house in the tree.
house - tree

Concept

Site Plan

Section

Facade pattern

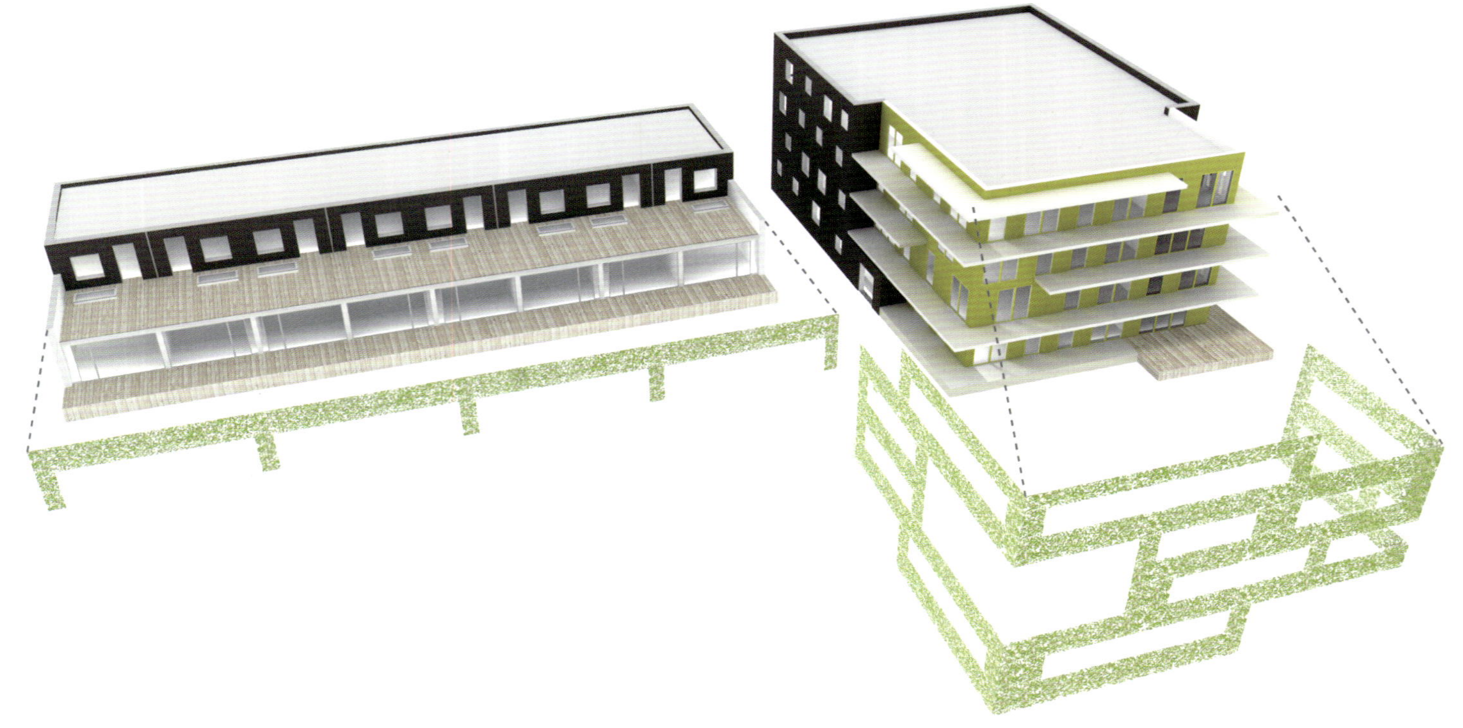

Axonometry

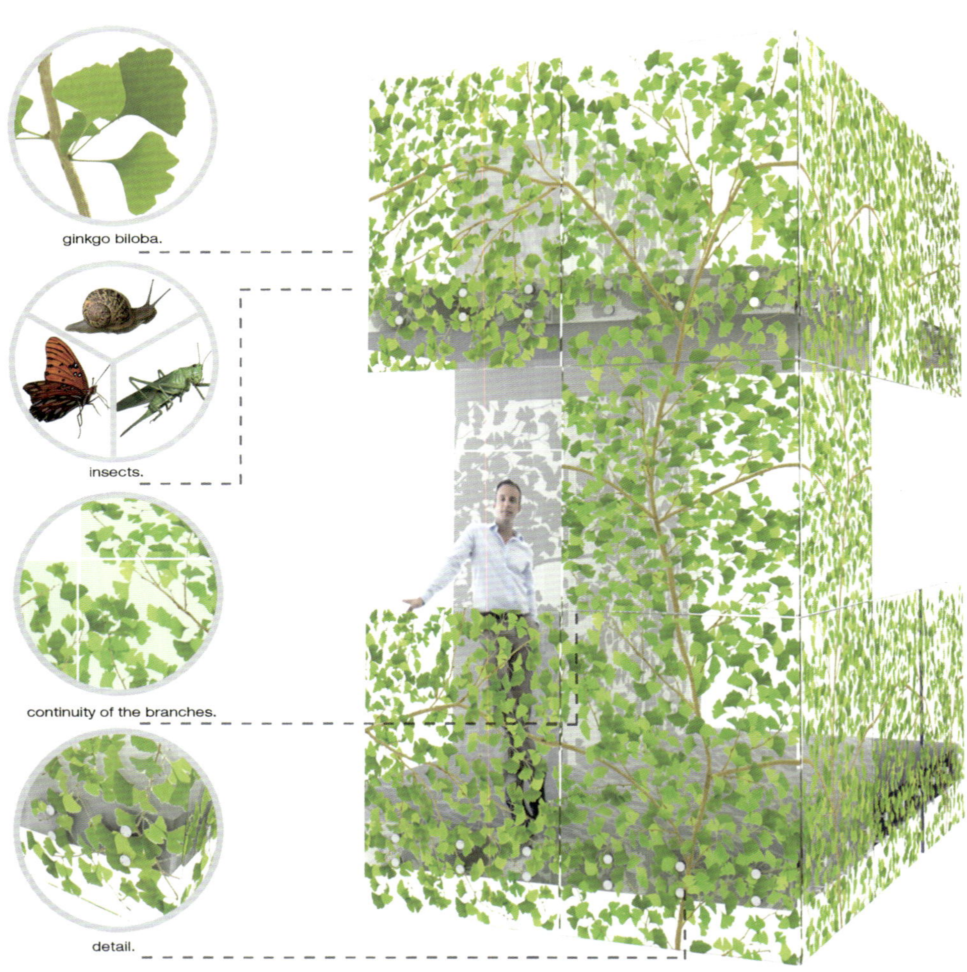

Glass 3d Scheme

Construction Process

© CHA

© CHA

© CHA

workx
© CHA

workx
© CHA

© CHA

© Christian Richters

© Christian Richters

First Floor Penthouse, Tel Aviv, Israel

Z-A studio

© Assaf Pinchuk

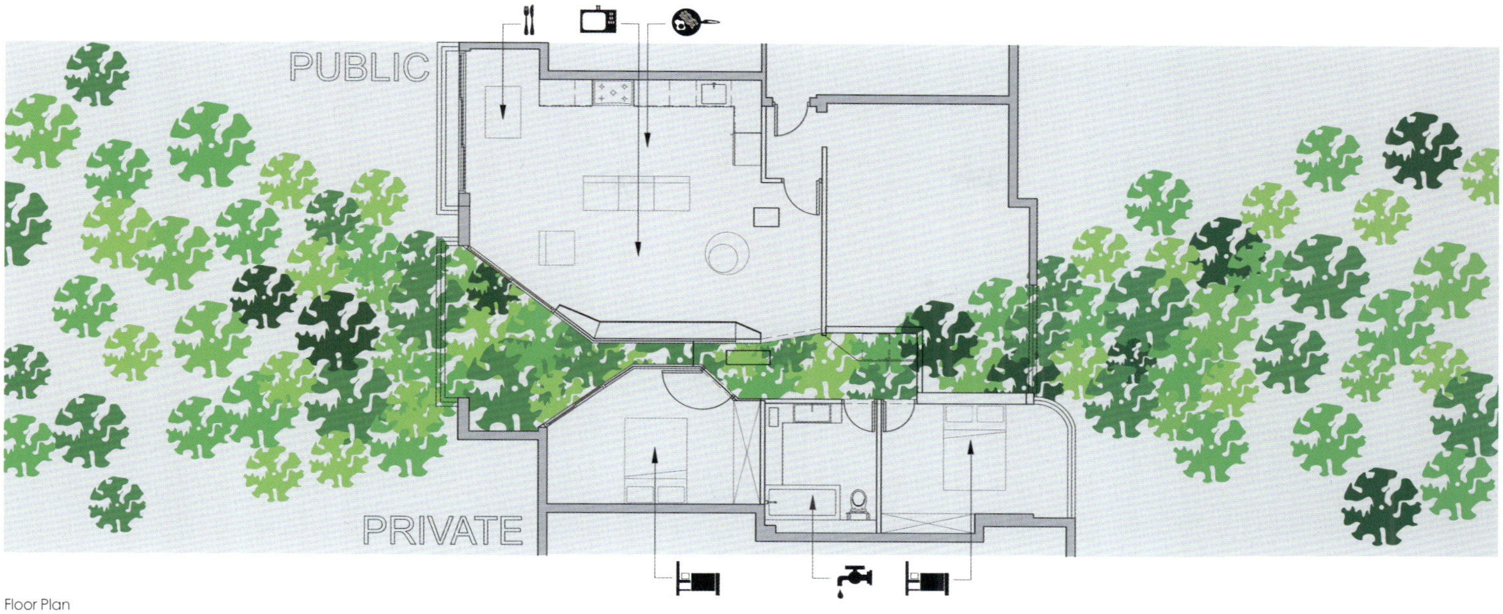

PUBLIC

PRIVATE

Floor Plan

Elevation

© Assaf Pinchuk

St. Nikolaus, Neumarkt, Austria
kadawittfeldarchitektur

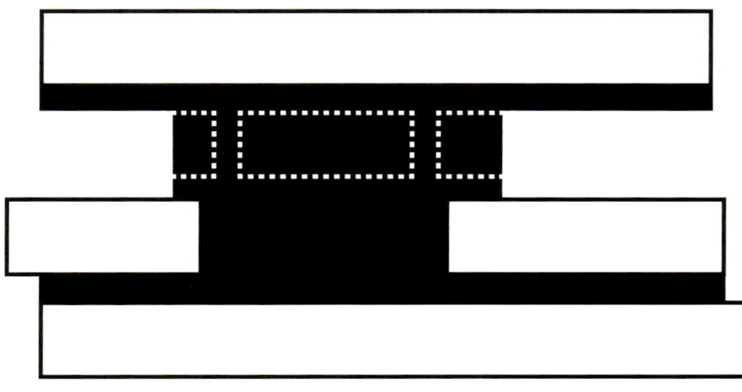
Conceptual Diagram

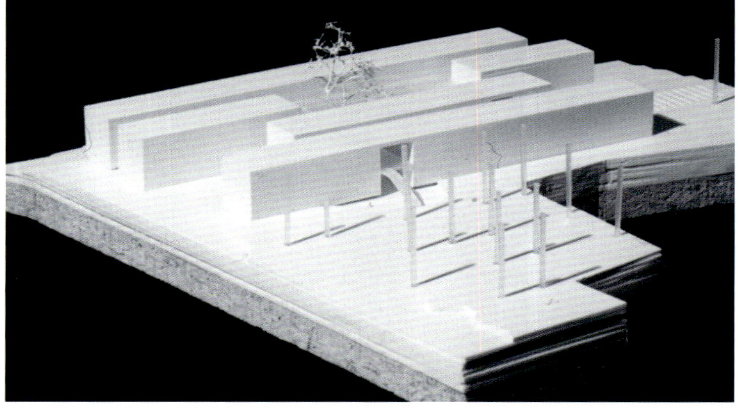
Volume Study Modeling

IM LABYRIN
VERLIERT M

SICH NICHT
SICH
WISSEN IST

VILLA TORPED, Saint Denis, France
PERIPHERIQUES

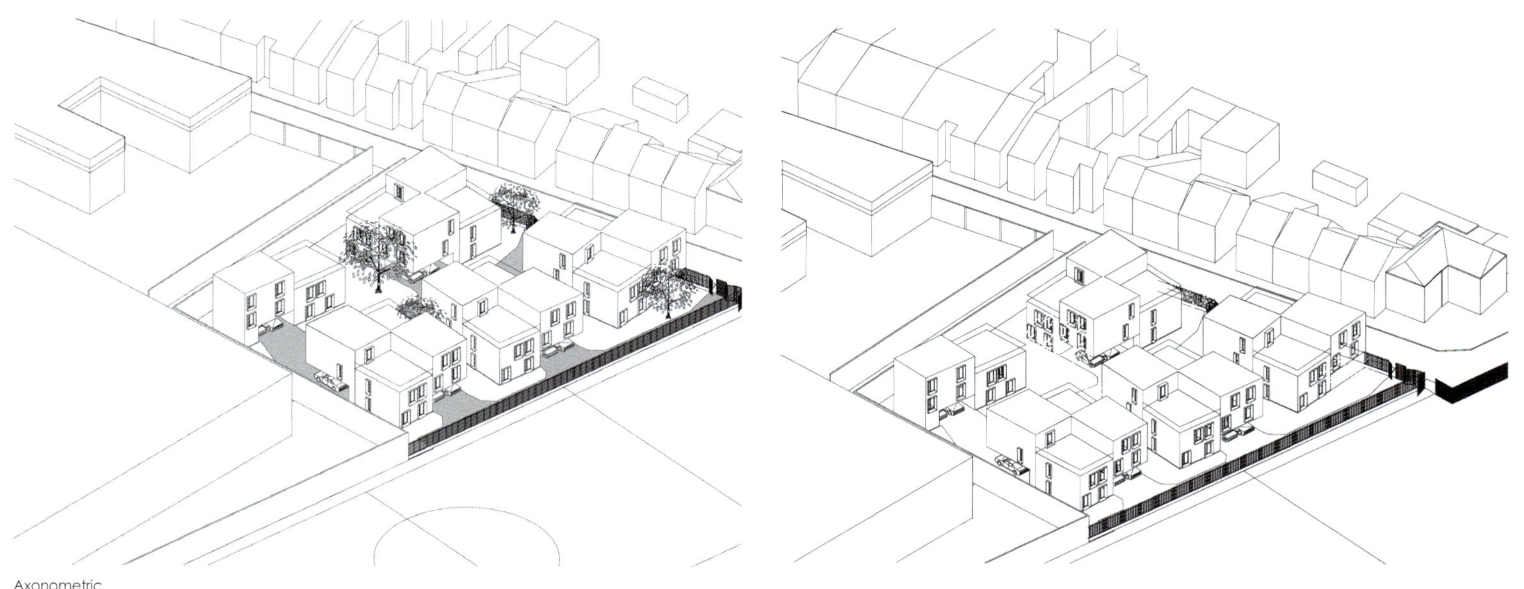

Axonometric

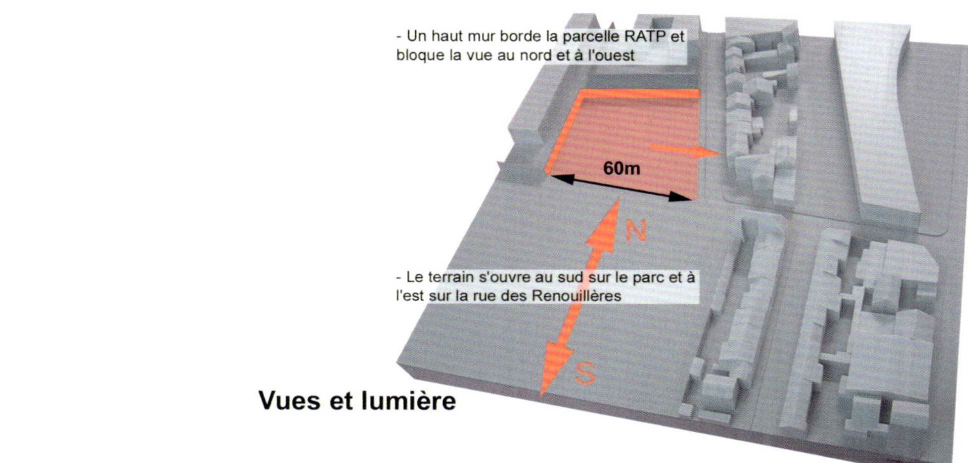

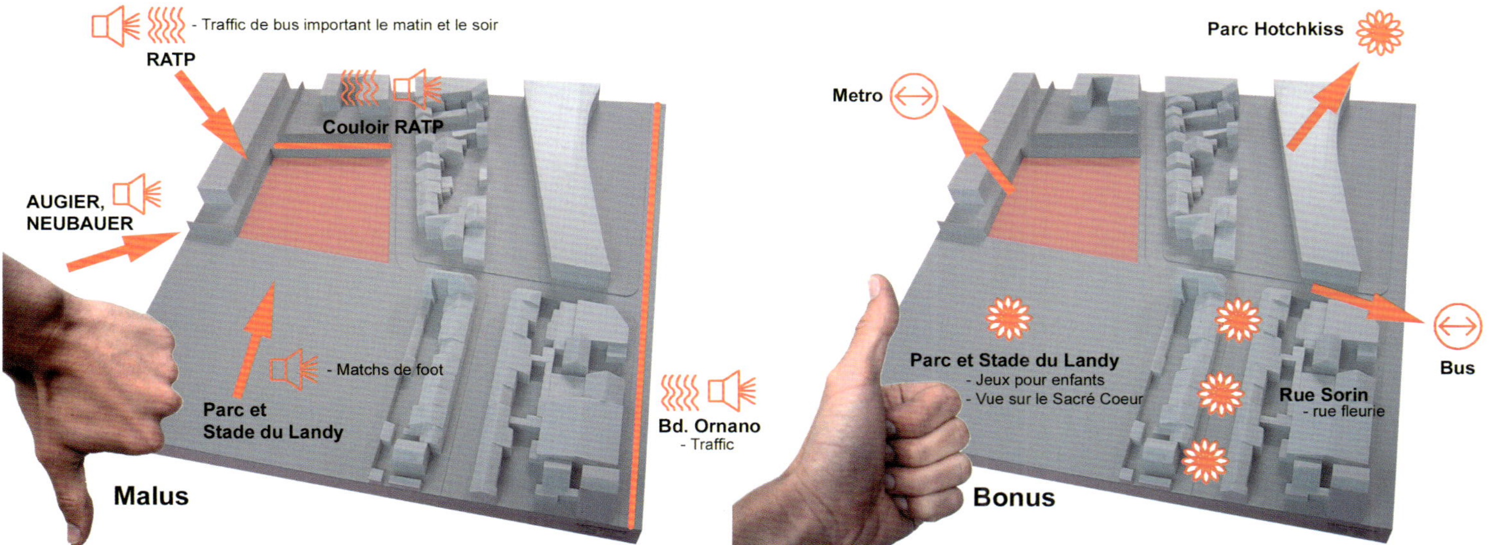

Site Analysis

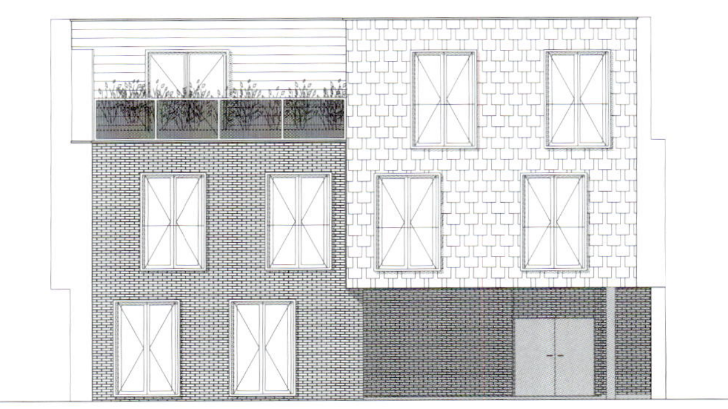

Elevation A3

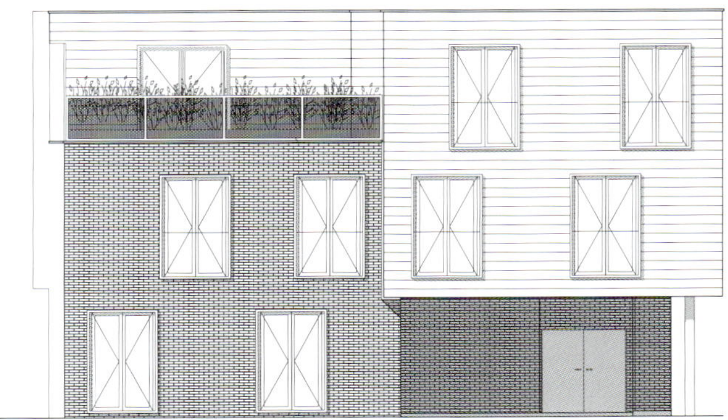

Elevation A1

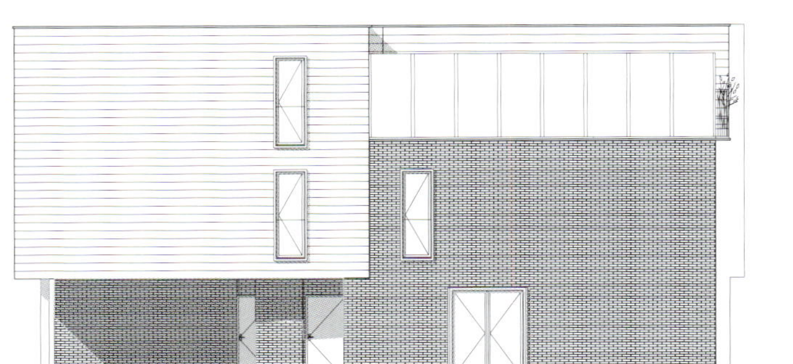

Elevation A2

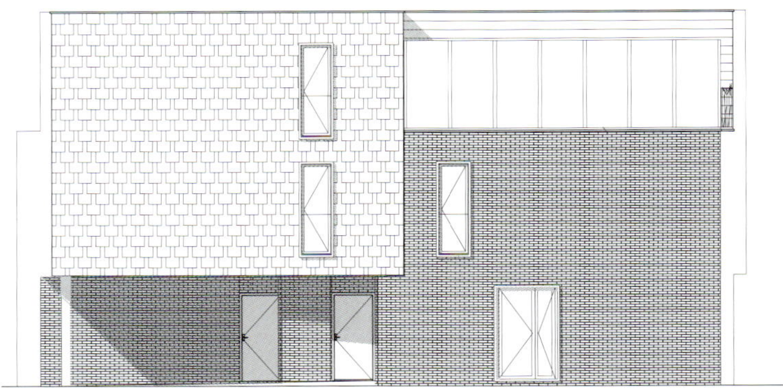

Elevation A4

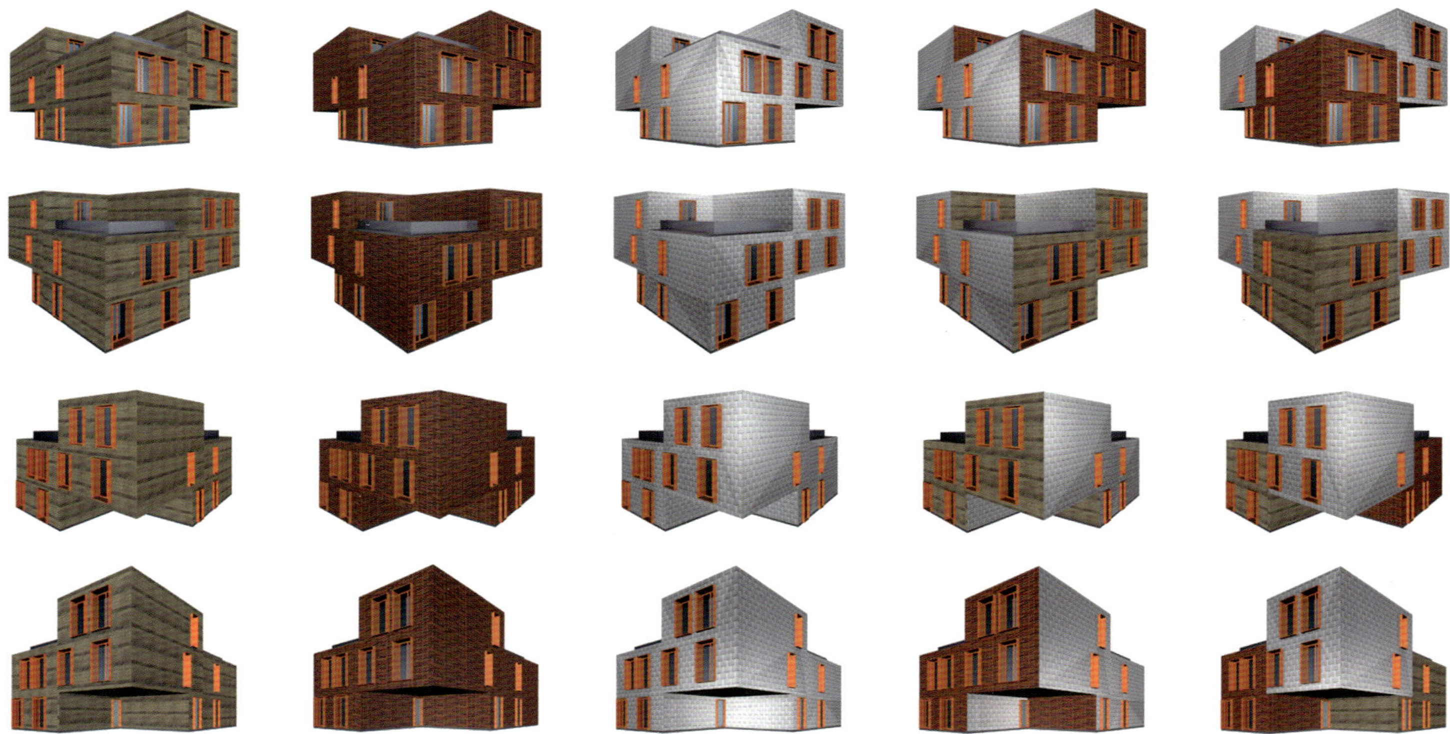

Collection of Volume Images

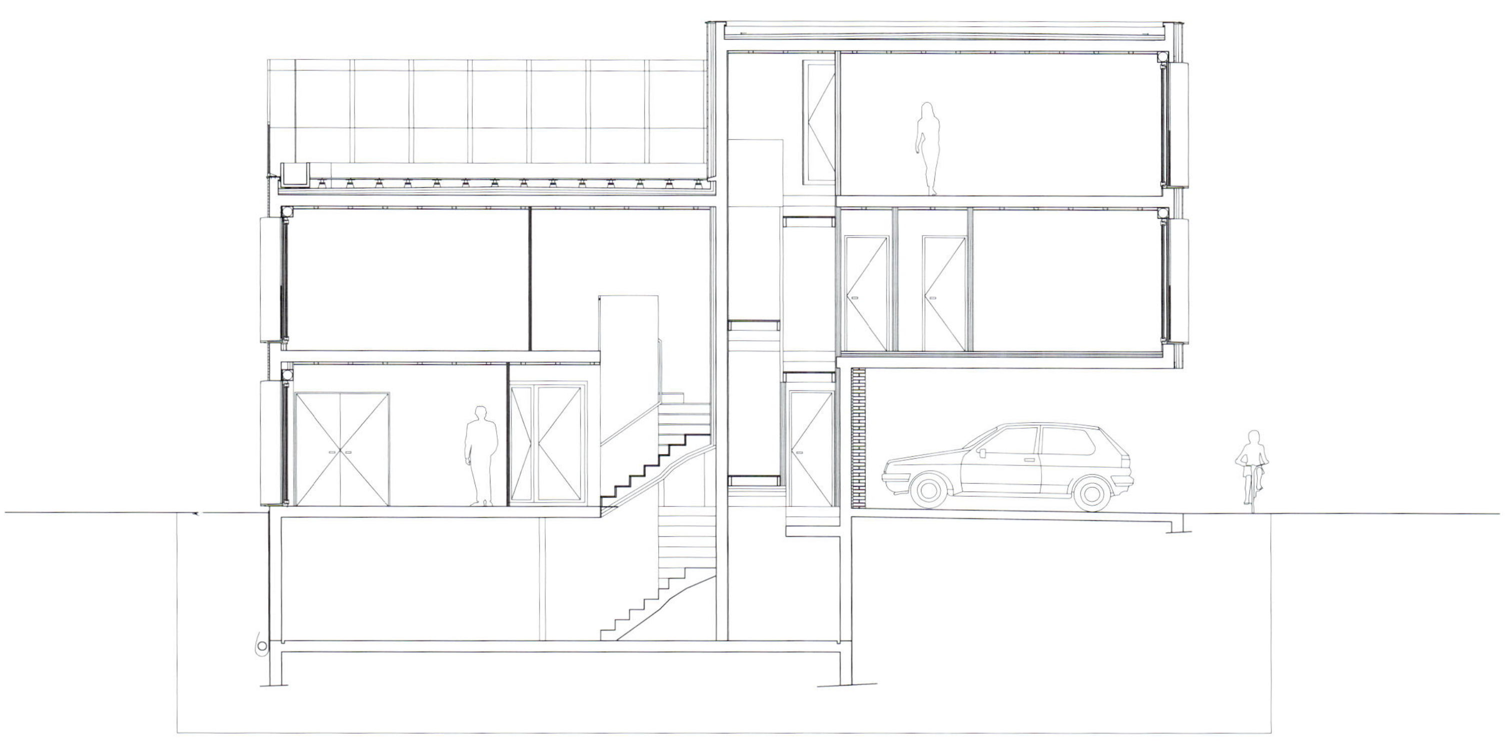

Section

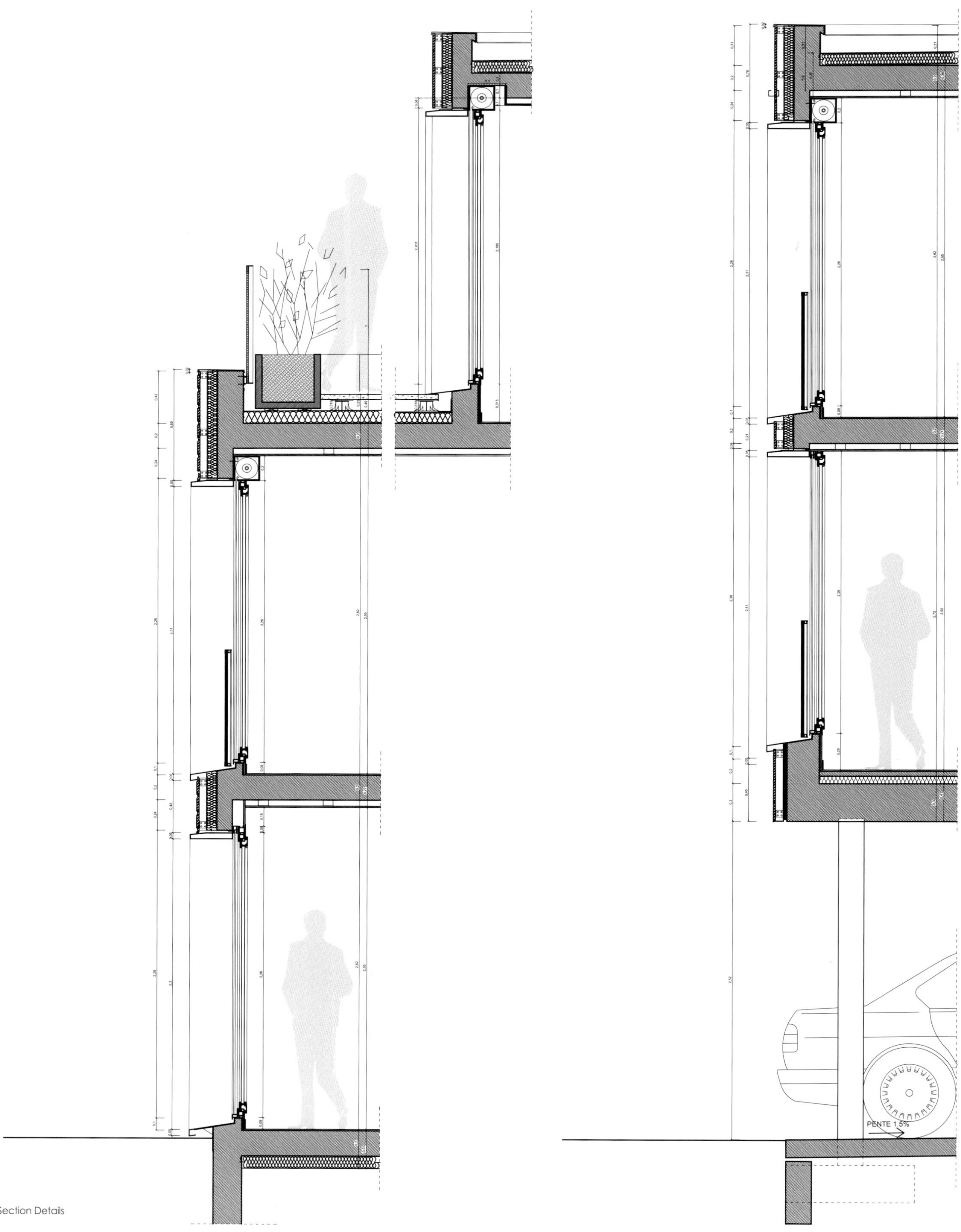

Section Details

VM Houses, Copenhagen, Denmark

BIG

Design Process

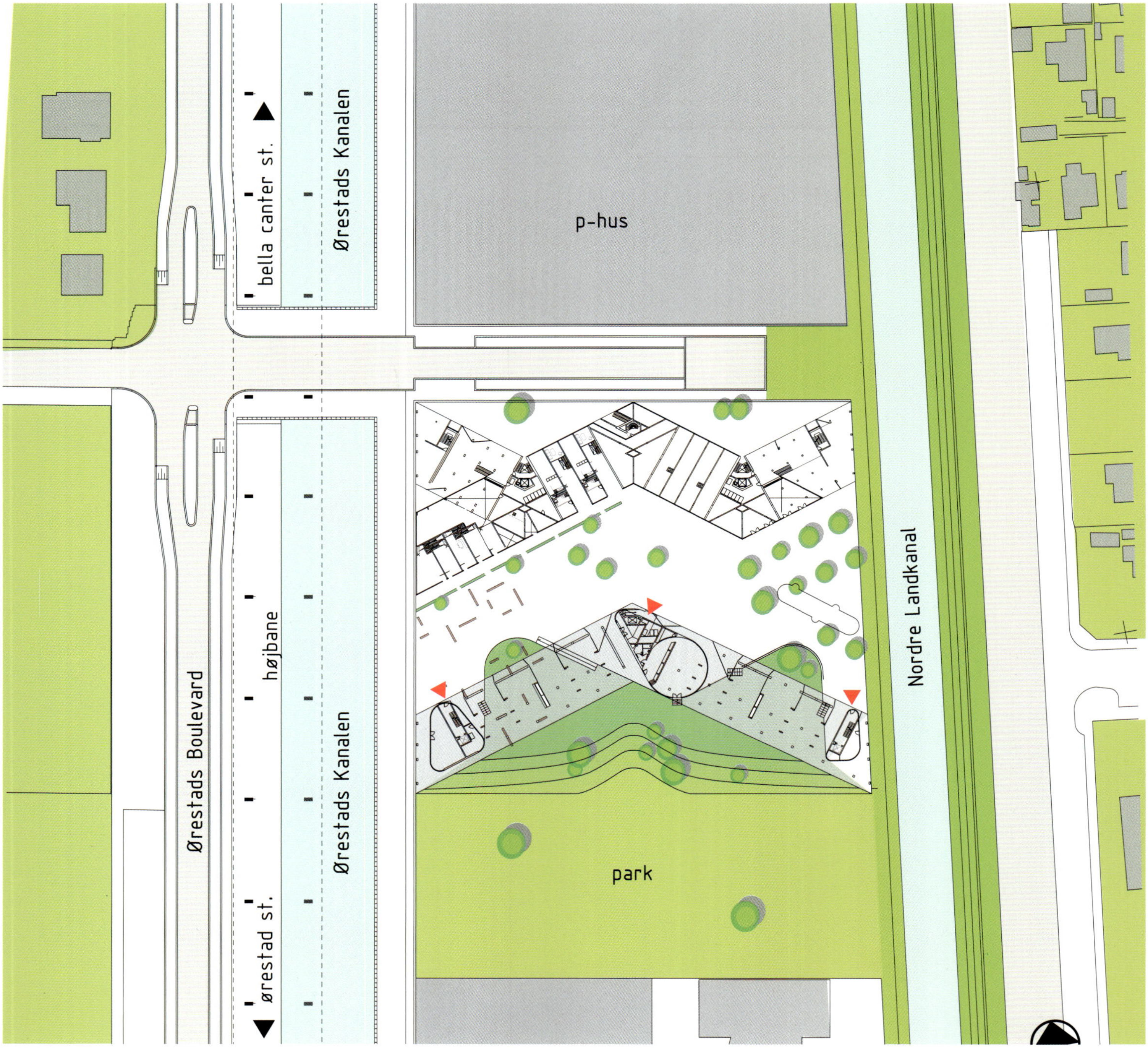

p-hus
bella canter st.
Ørestads Kanalen
højbane
Ørestads Boulevard
Ørestads Kanalen
ørestad st.
Nordre Landkanal
park

Ground Floor

Fourth Floor

© Nicholai Moeller

© Nicholai Moeller

© BIG

© Peter Boel

© Jasper Carlberg

© Peter Boel

Cyprus Medical School
object-e architecture

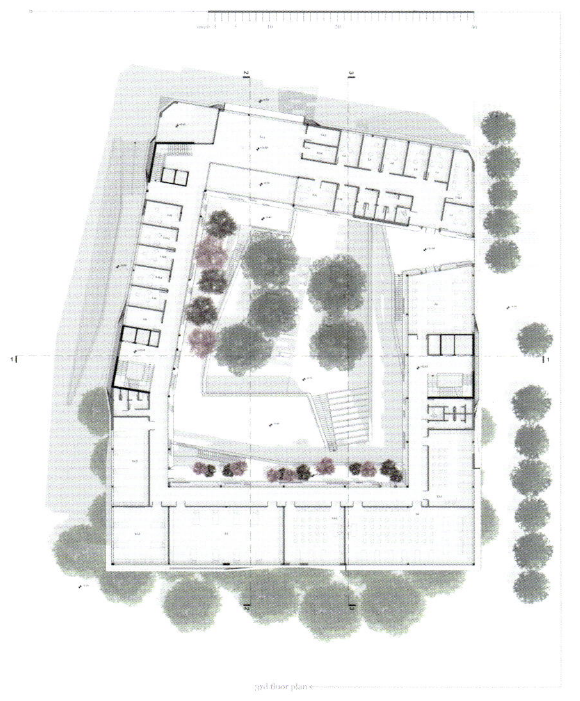

Site Plan

Study Modeling

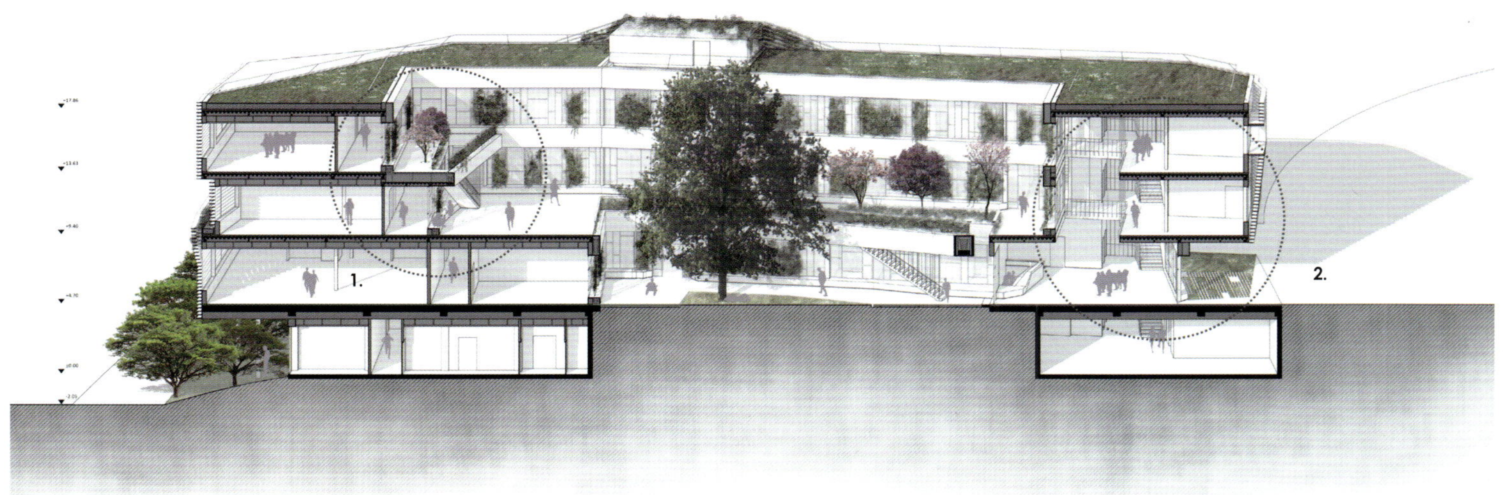

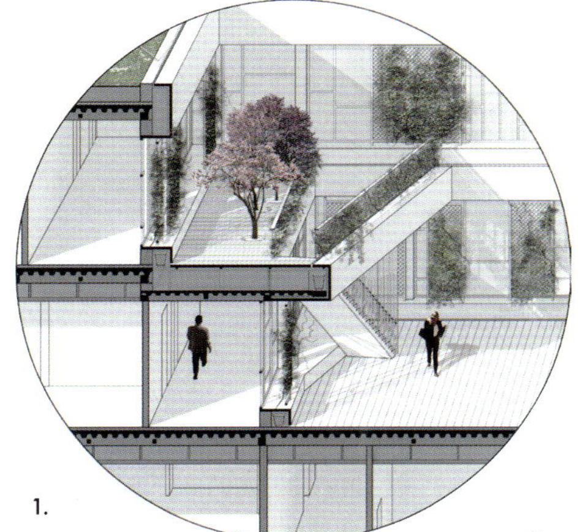

Section 2-2

1. Zoom Sxedio 1
2. Zoom Sxedio 2

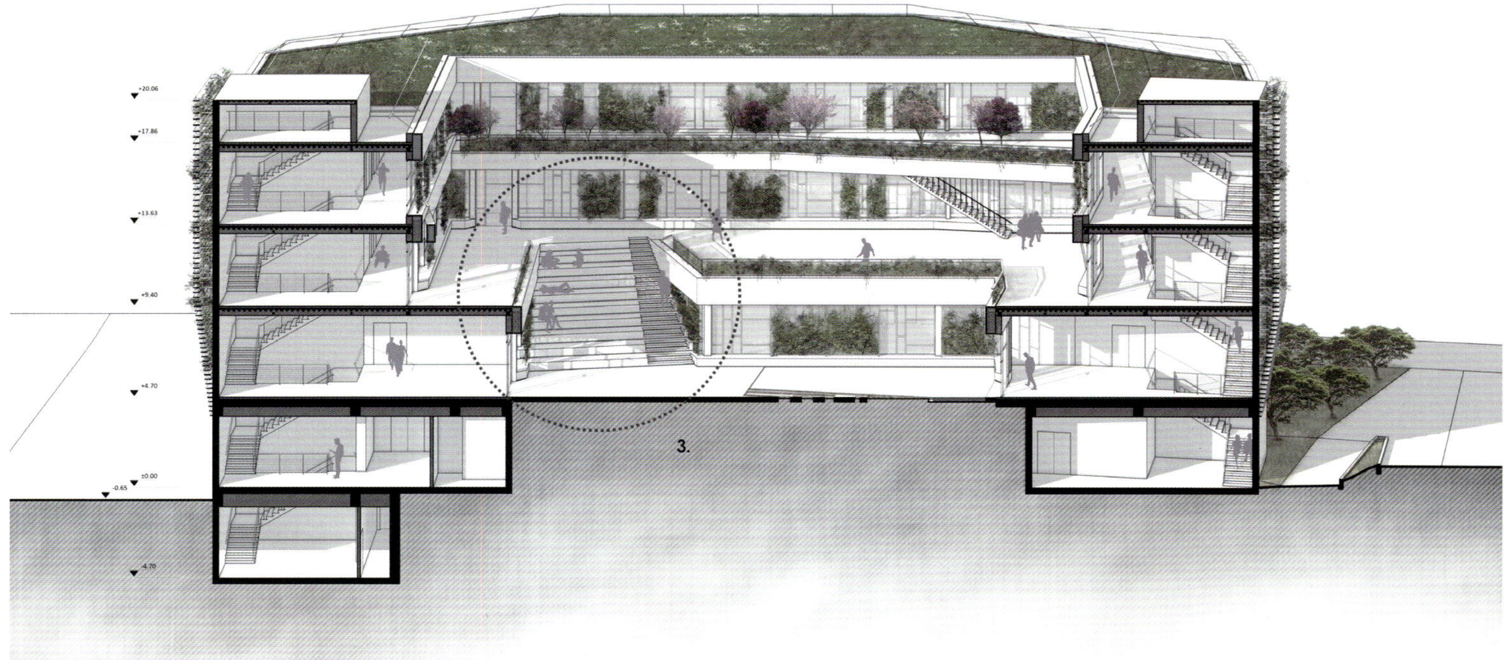

Section 1-1

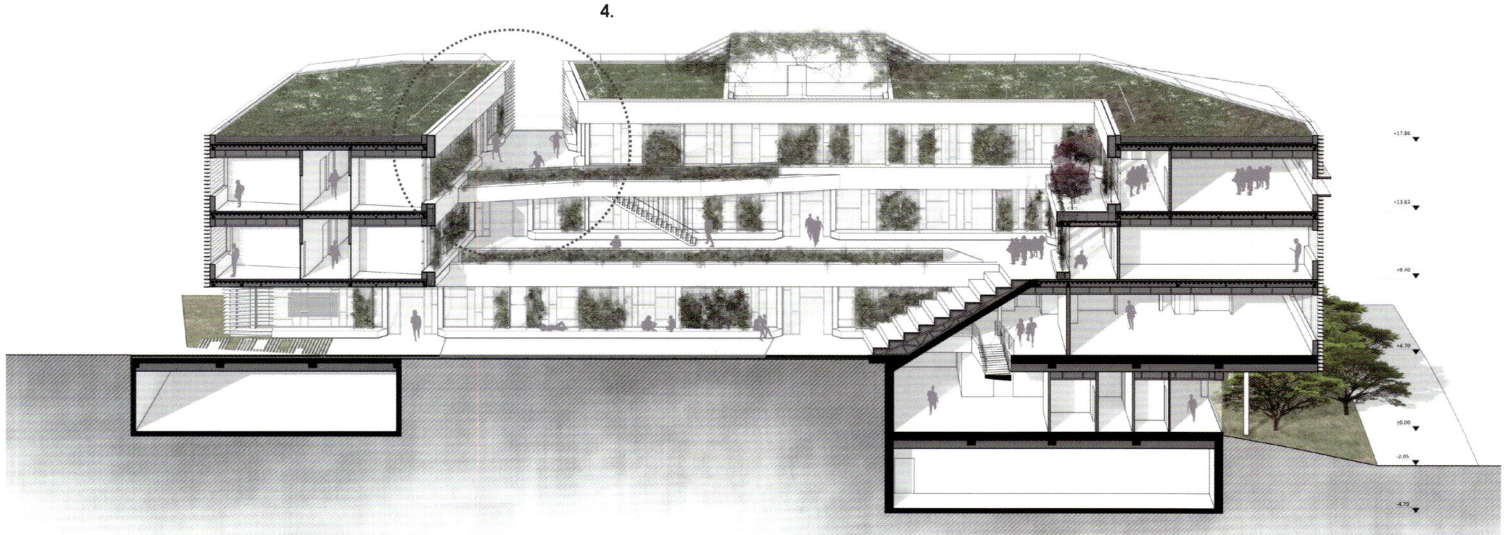

Section 3-3

Section

1. Zoom Sxedio 3
2. Zoom Sxedio 4

Grundfos Halls of Residence, Aarhus, Denmark

CEBRA

© Mikkel Frost | CEBRA

© Mikkel Frost | CEBRA

© Mikkel Frost | CEBRA

© Mikkel Frost | CEBRA

© Mikkel Frost | CEBRA

07.11
06.11
03.11

© Mikkel Frost | CEBRA

School at Kuchl, Kuchl, Austria

kadawittfeldarchitektur

Open Space

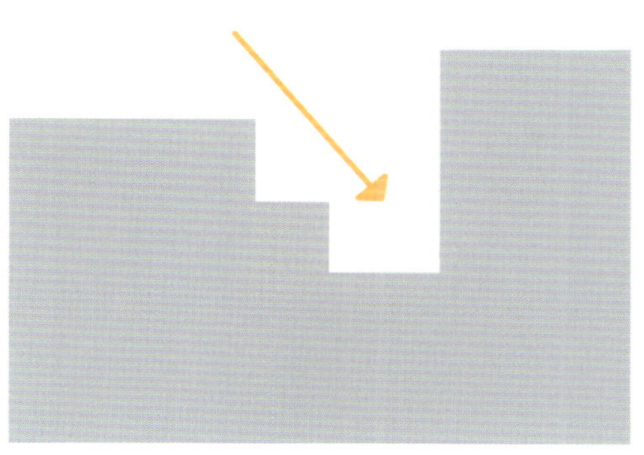
Cubature

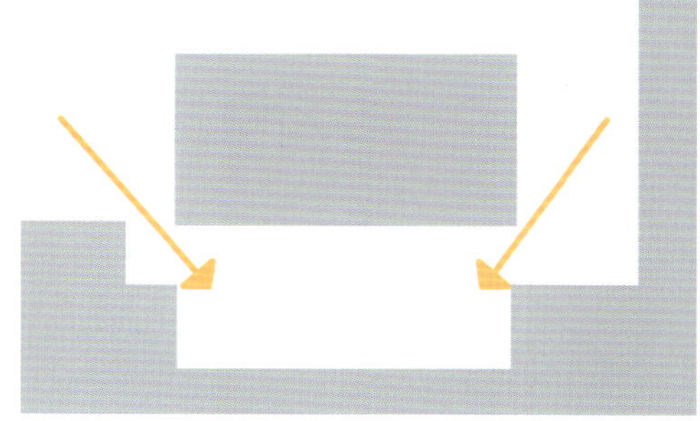
Daylight

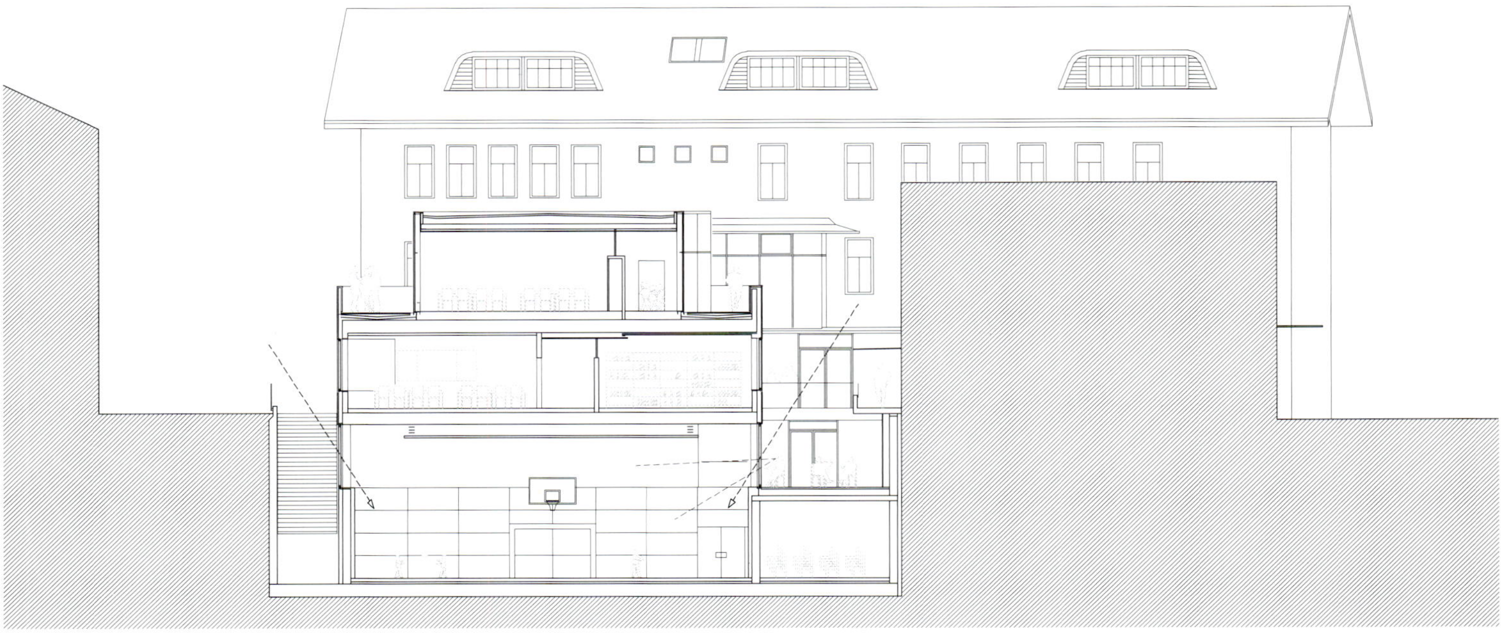

Section A-A

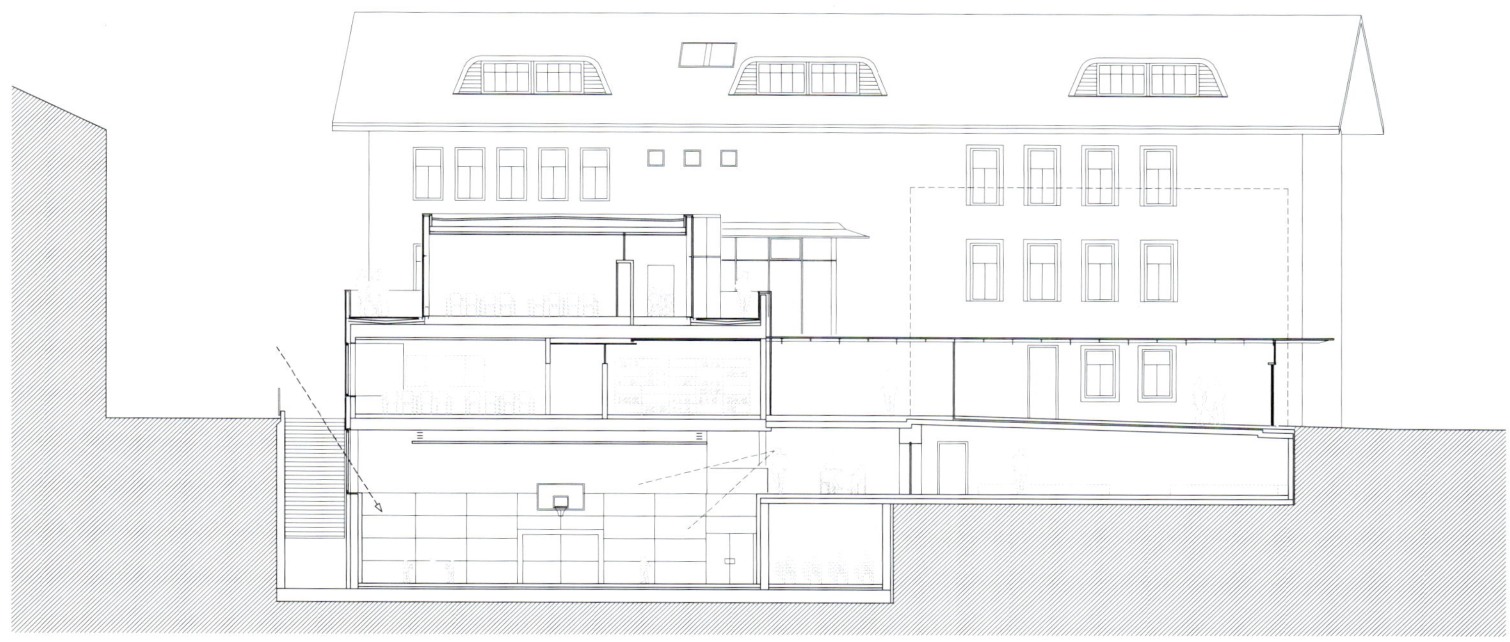

Section B-B

School at Mittersill, Mittersill, Austria

kadawittfeldarchitektur

Existing Building Structure ...

... Metamorphosis ...

... Transformation into a new Building

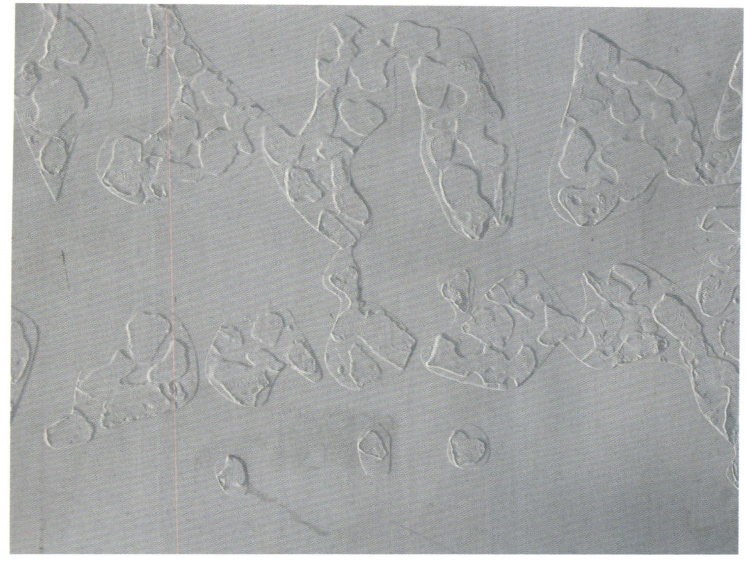

Mold of Concrete Wall Base

TURKNA

University Campus & Research Park, Salzburg-Urstein, Austria

kadawittfeldarchitektur

Site Plan

1. Historical Dairy-Farm /2. Castle & Pavillon
3. University / 4. Campus

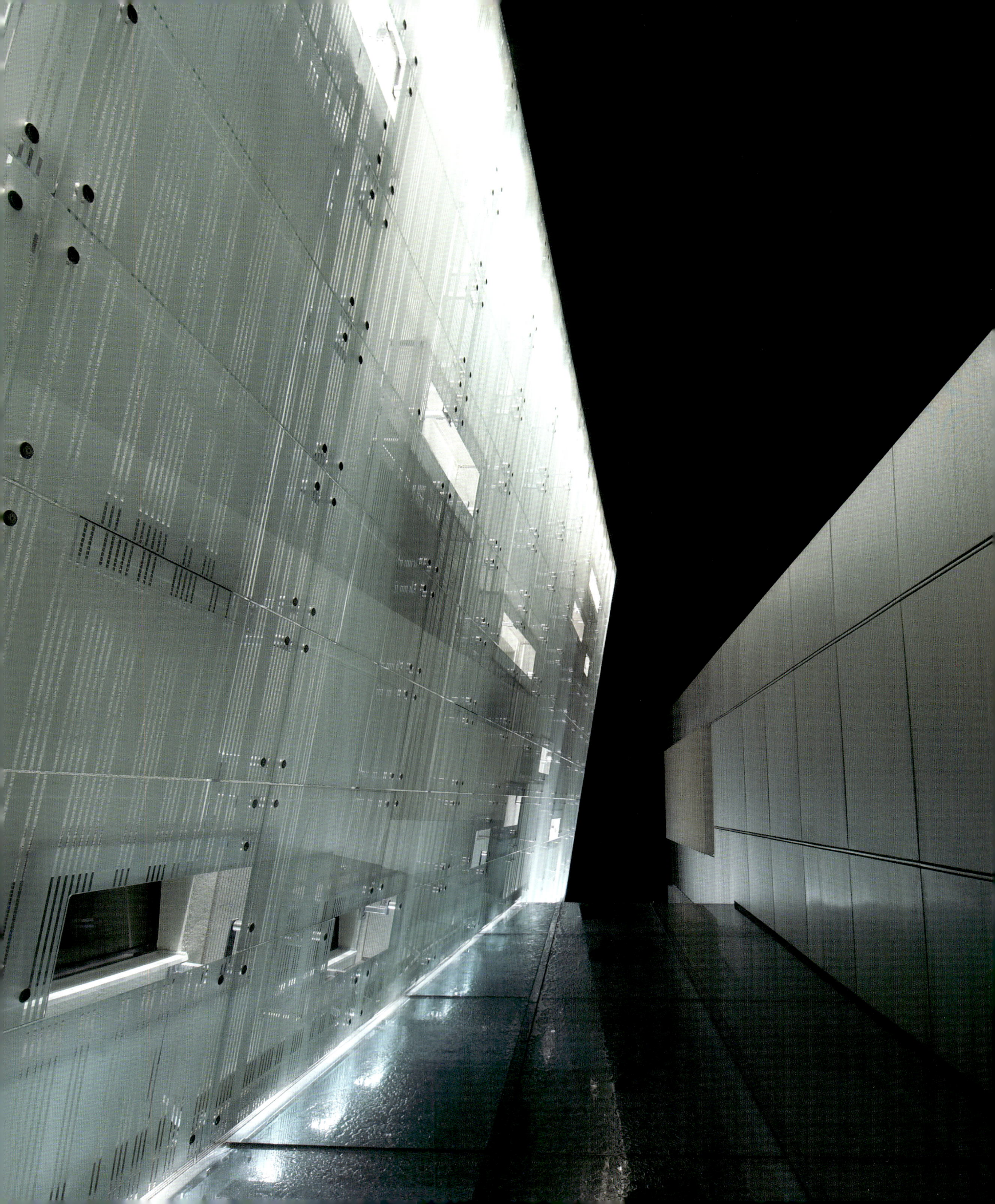

Elavation Detail

University of Southern Denmark, Kolding, Denmark

Henning Larsen Architects

© Hufton + Crow

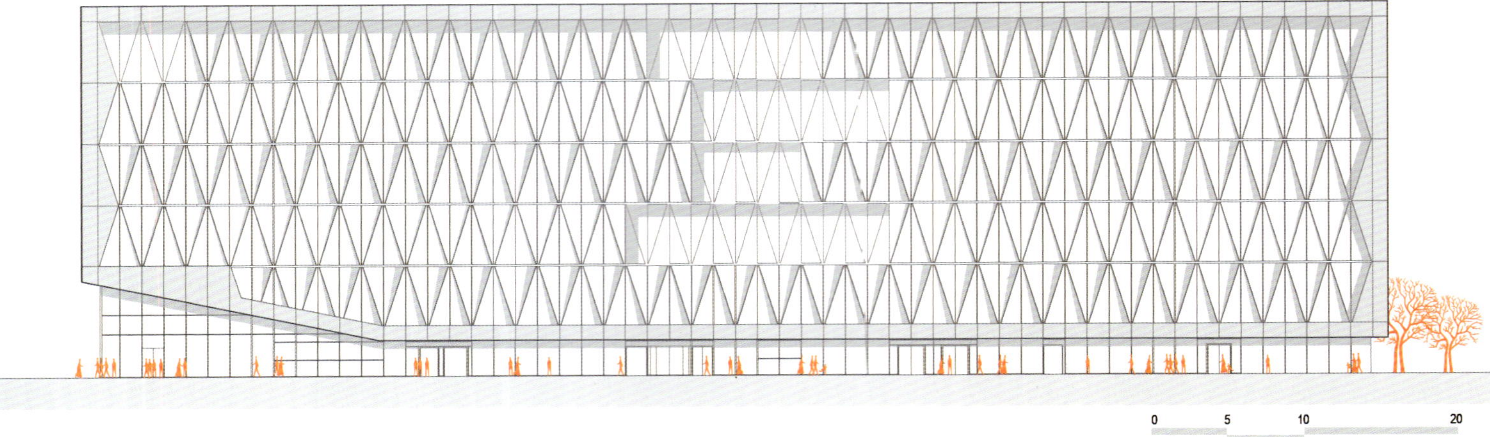

Elevation A

© Hufton + Crow

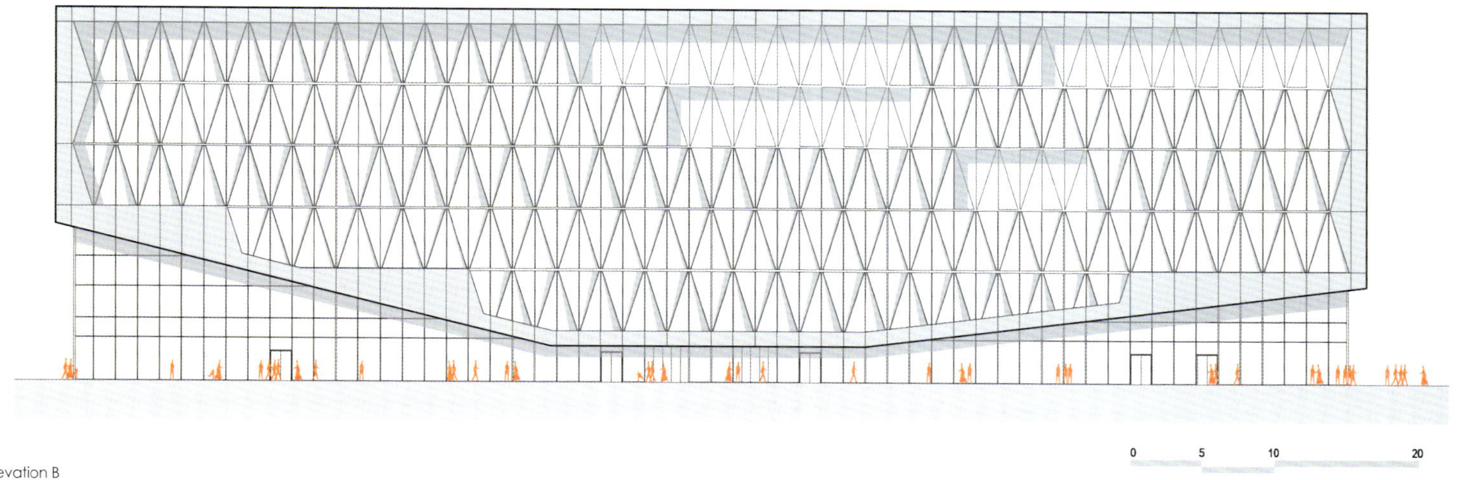

Elevation B

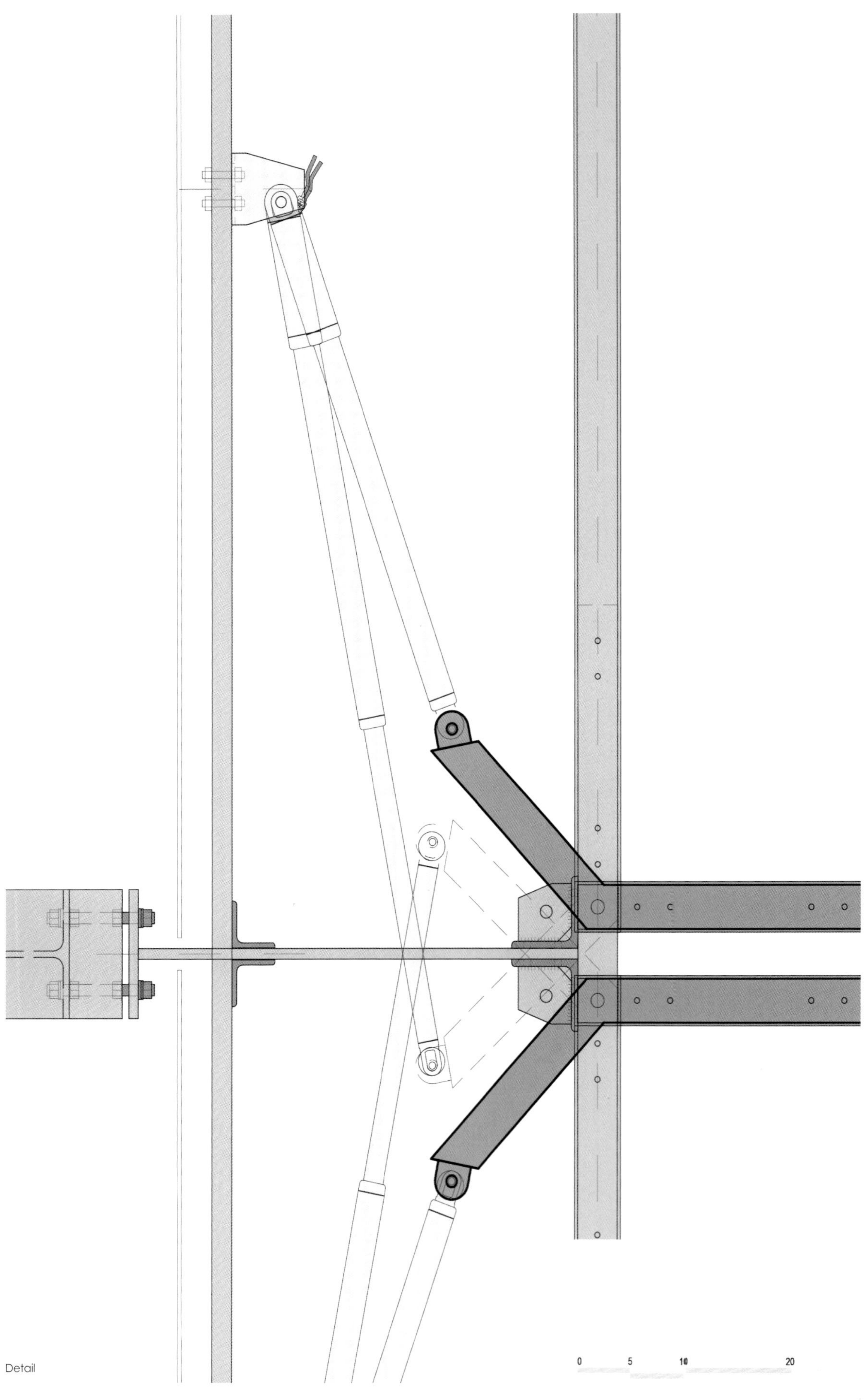

Detail

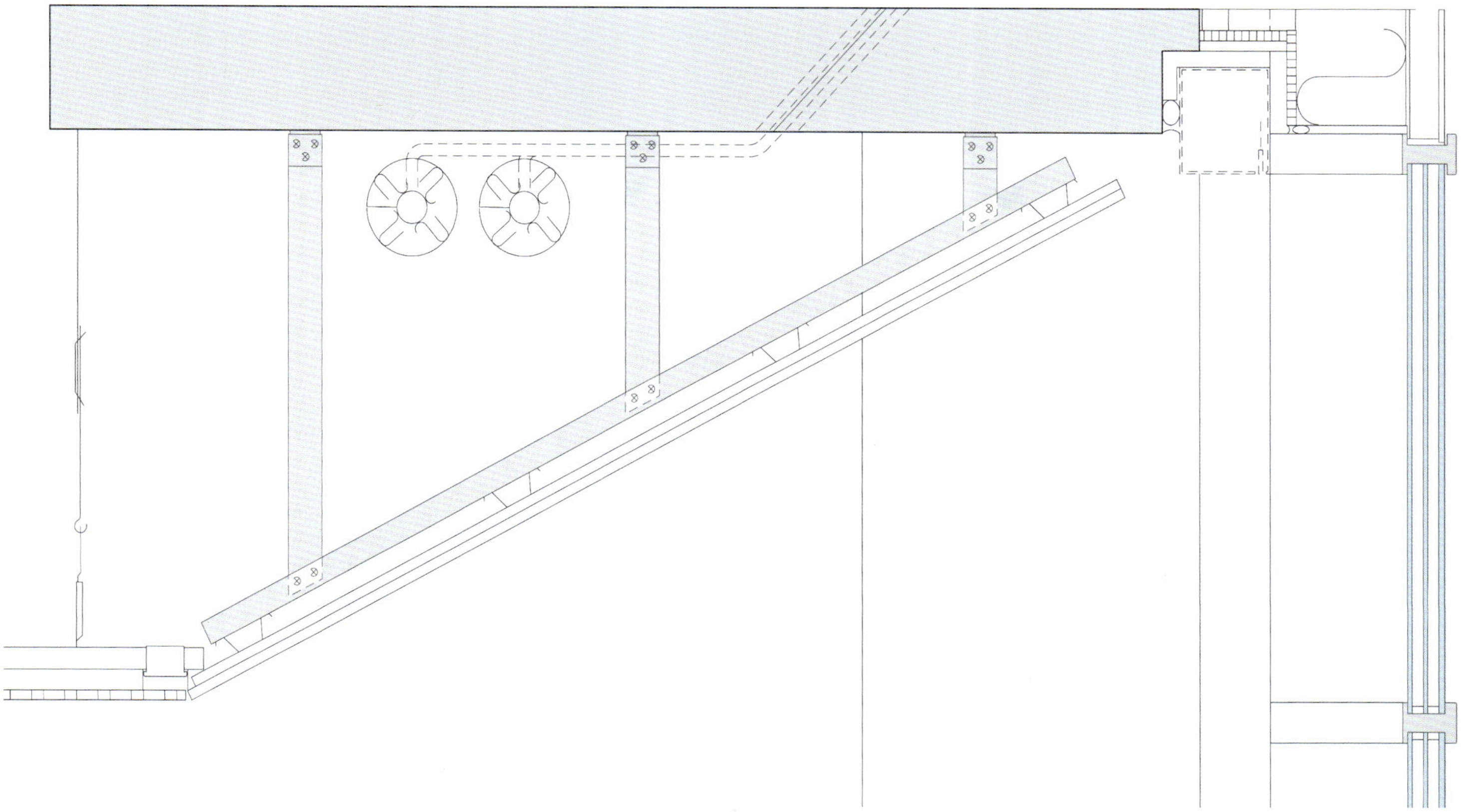

Detail

745 Seventh Avenue, New York, U.S.A

Kevin Kennon Architects

PM
ANKFURT
LONDON
NEW YORK
brandy
the new album
available at
fye
never stop playing
music • movies • games • more
brandy
brandy
HOULIHAN'S
RESTAURANT AND BAR
HOULIHAN'S
RESTAURANT
AND BAR
GOURMET DELI
STAND

adidas

OFFICE TRADING
TRADIN

adidas

NEW YORK 06:37
LONDON 02:37:21 AM
RETAIL
RETAIL
SUBWAY
SUBWAY
HOULIHAN'S
RESTAURANT
AND BAR
Tex-Mex Restaurant
and Cantina

Beijing Conrad Hotel, Beijing, China

MAD Architects

© XiaZhi

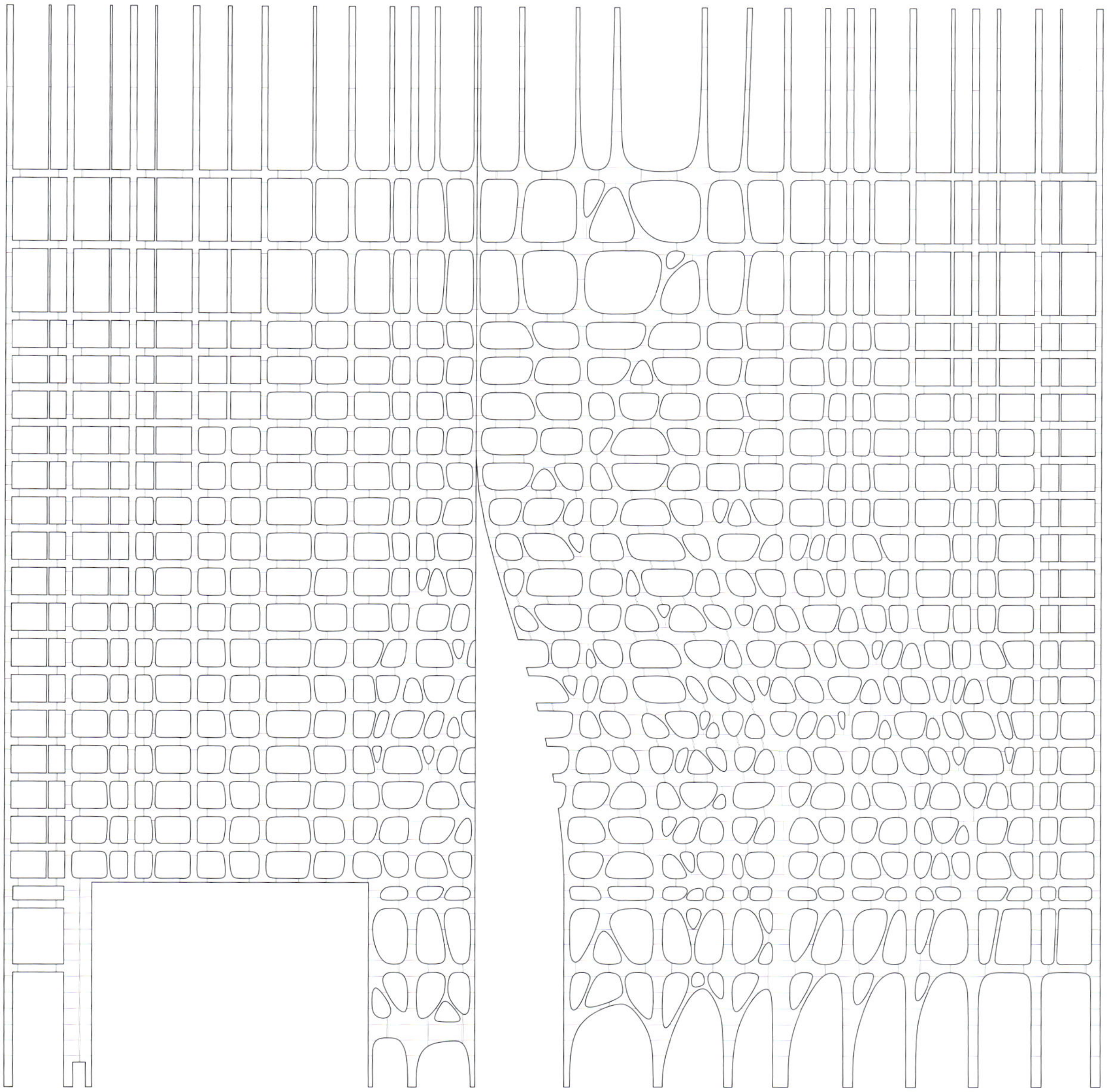

Renderings + Alum

© XiaZhi

© XiaZhi

© XiaZhi

© XiaZhi

Box in the Box, Madrid, Spain
Arenas Basabe Palacios Arquitectos
© Imagen Subliminal

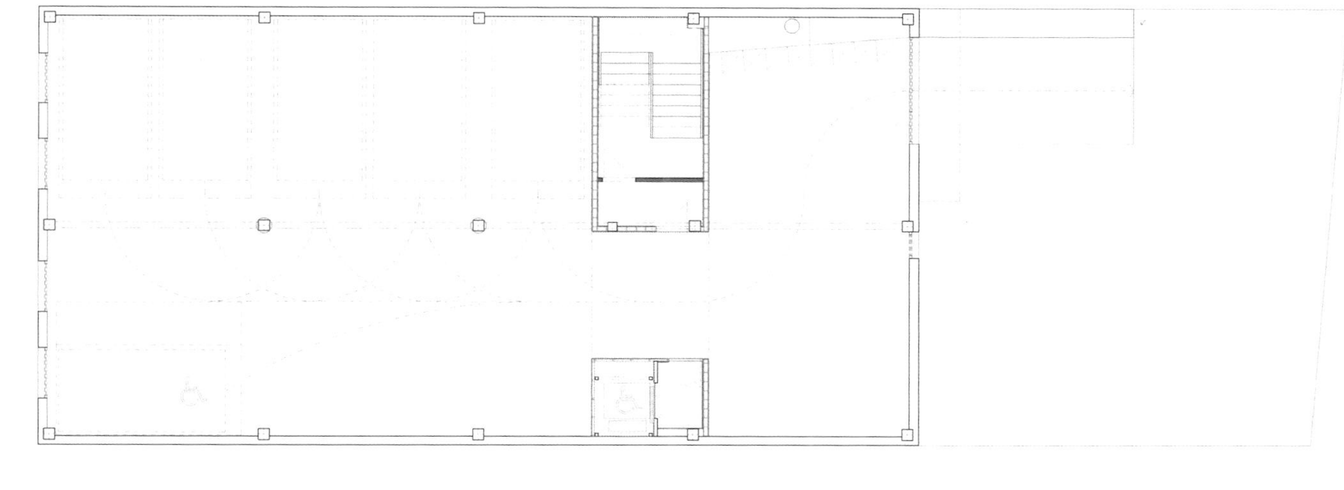

Basement Floor

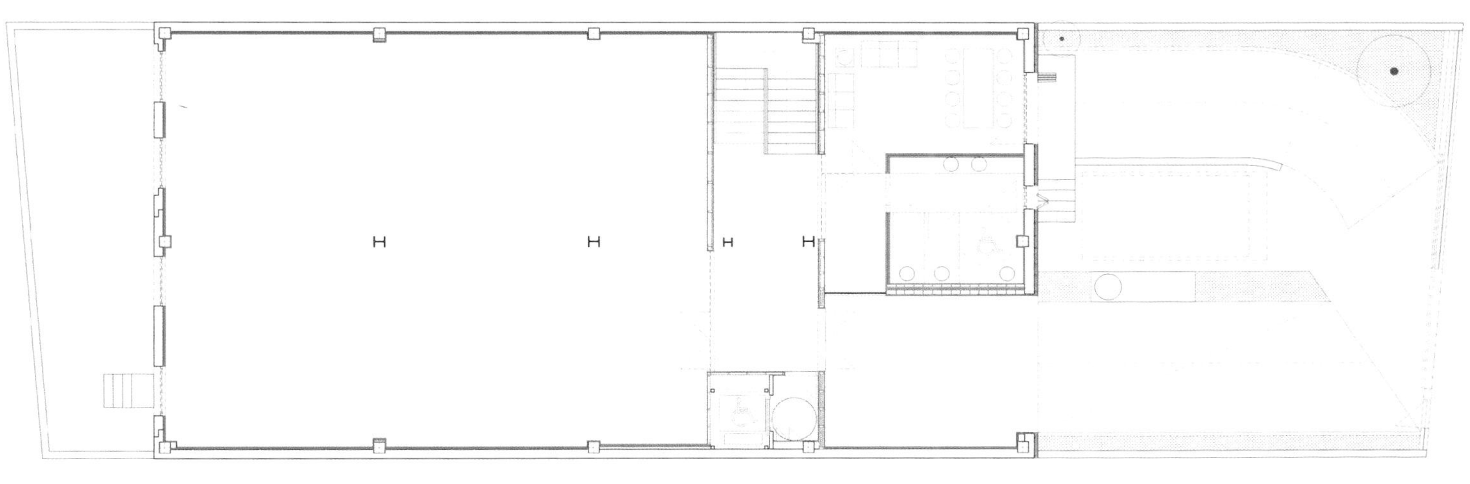

Ground Floor

Elevation A-A

Elevation B-B

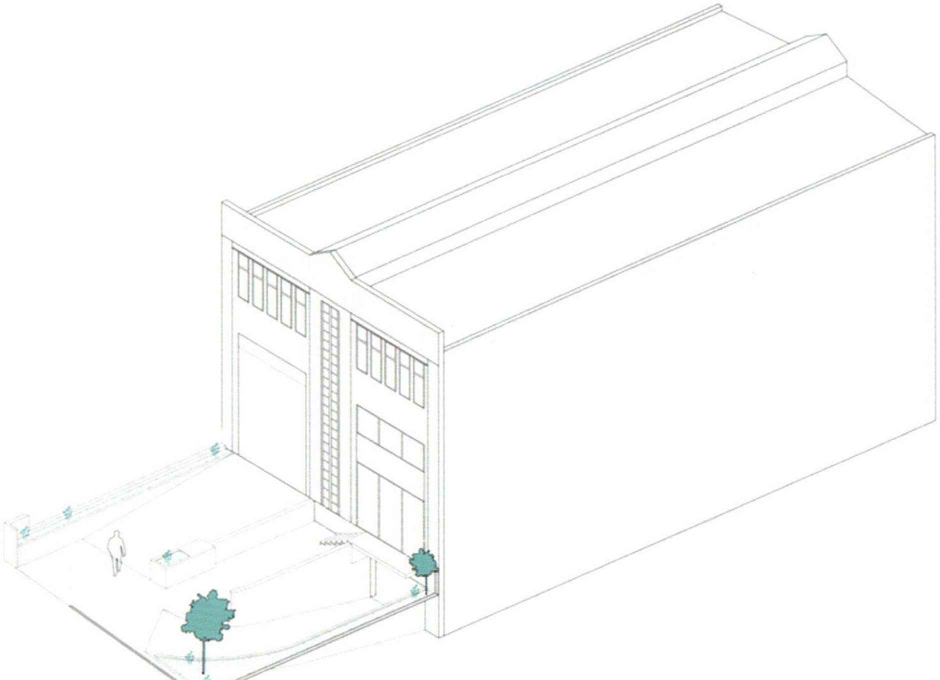

Axnometirc (Before)

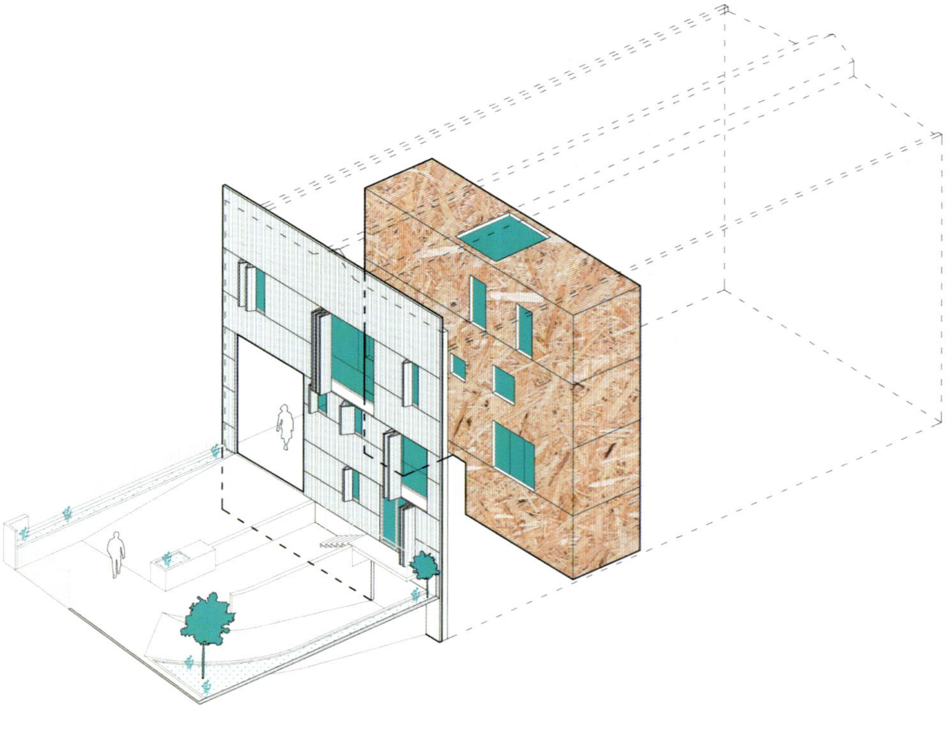

Axnometirc (After)

© Imagen Subliminal

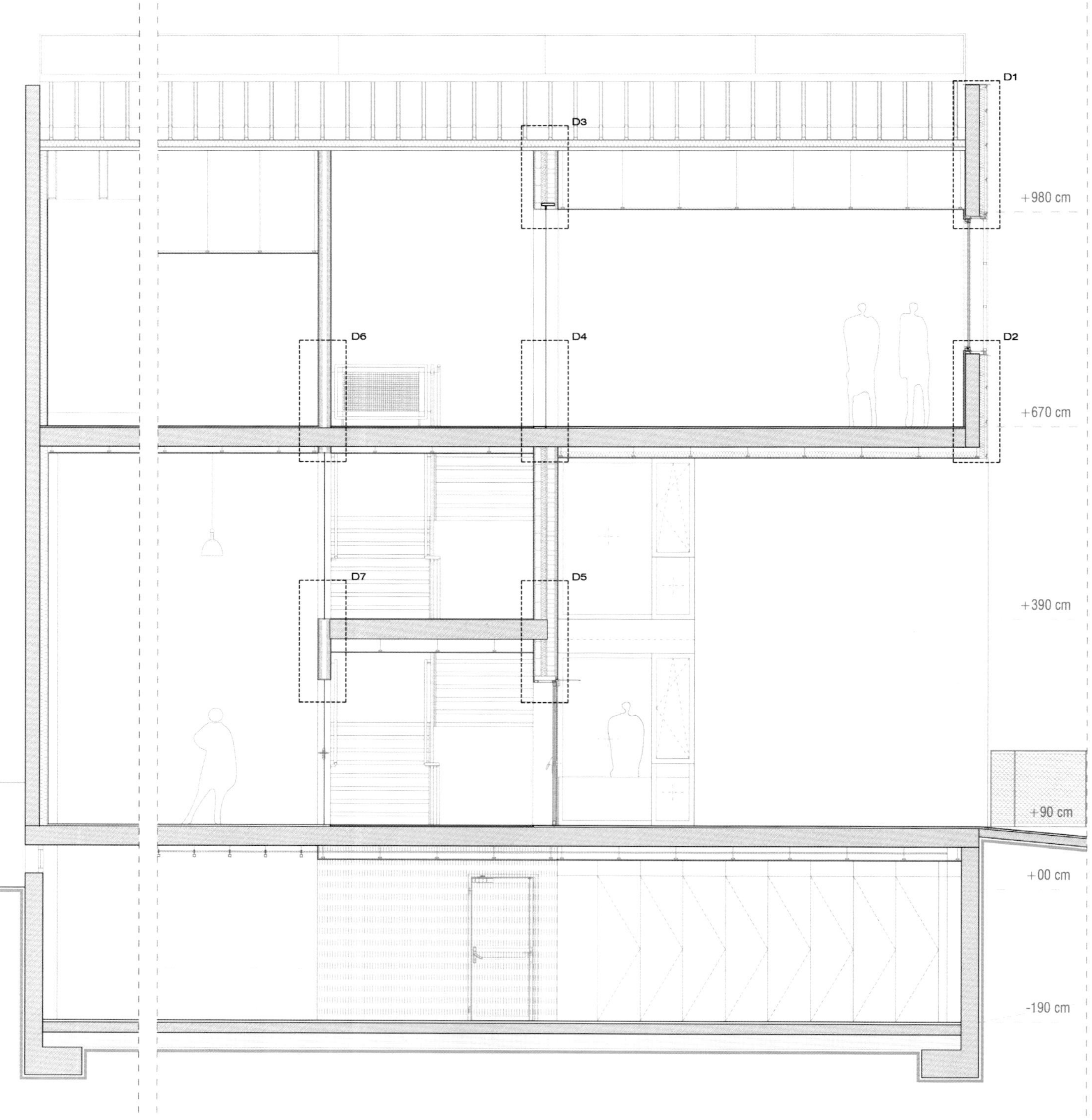

Section 1:50

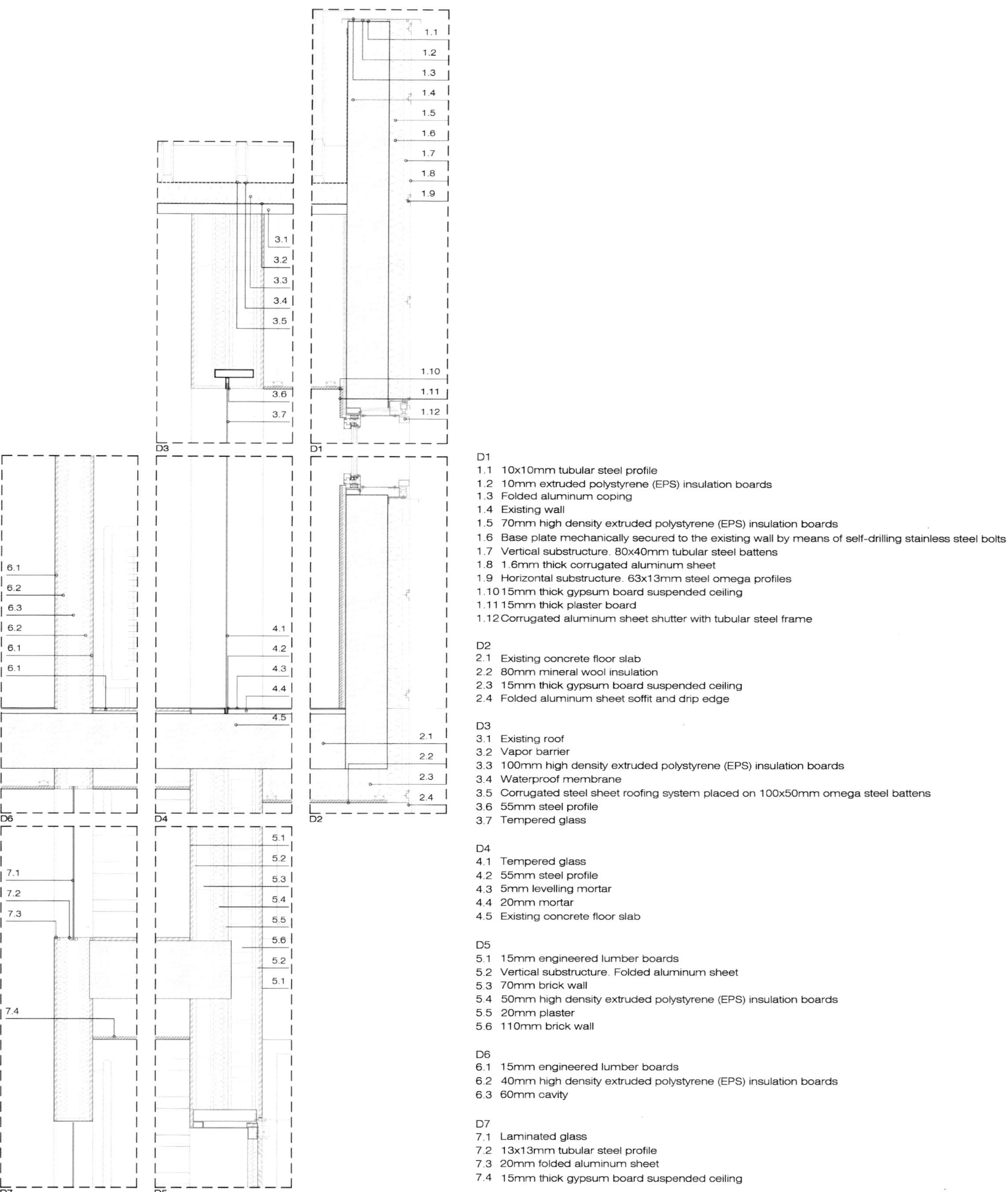

D1
1.1 10x10mm tubular steel profile
1.2 10mm extruded polystyrene (EPS) insulation boards
1.3 Folded aluminum coping
1.4 Existing wall
1.5 70mm high density extruded polystyrene (EPS) insulation boards
1.6 Base plate mechanically secured to the existing wall by means of self-drilling stainless steel bolts
1.7 Vertical substructure. 80x40mm tubular steel battens
1.8 1.6mm thick corrugated aluminum sheet
1.9 Horizontal substructure. 63x13mm steel omega profiles
1.10 15mm thick gypsum board suspended ceiling
1.11 15mm thick plaster board
1.12 Corrugated aluminum sheet shutter with tubular steel frame

D2
2.1 Existing concrete floor slab
2.2 80mm mineral wool insulation
2.3 15mm thick gypsum board suspended ceiling
2.4 Folded aluminum sheet soffit and drip edge

D3
3.1 Existing roof
3.2 Vapor barrier
3.3 100mm high density extruded polystyrene (EPS) insulation boards
3.4 Waterproof membrane
3.5 Corrugated steel sheet roofing system placed on 100x50mm omega steel battens
3.6 55mm steel profile
3.7 Tempered glass

D4
4.1 Tempered glass
4.2 55mm steel profile
4.3 5mm levelling mortar
4.4 20mm mortar
4.5 Existing concrete floor slab

D5
5.1 15mm engineered lumber boards
5.2 Vertical substructure. Folded aluminum sheet
5.3 70mm brick wall
5.4 50mm high density extruded polystyrene (EPS) insulation boards
5.5 20mm plaster
5.6 110mm brick wall

D6
6.1 15mm engineered lumber boards
6.2 40mm high density extruded polystyrene (EPS) insulation boards
6.3 60mm cavity

D7
7.1 Laminated glass
7.2 13x13mm tubular steel profile
7.3 20mm folded aluminum sheet
7.4 15mm thick gypsum board suspended ceiling

Section Detail 1:15

© Imagen Subliminal

© Imagen Subliminal

Chamber of Commerece and Industry, Ljubljana, Slovenia

Sadar Vuga Architekti

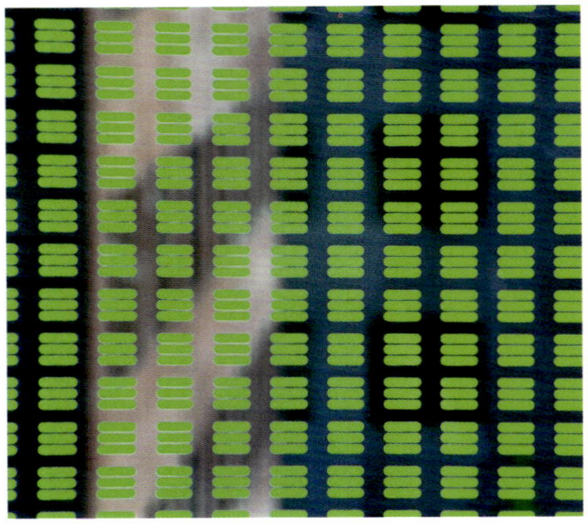

Solar Panels

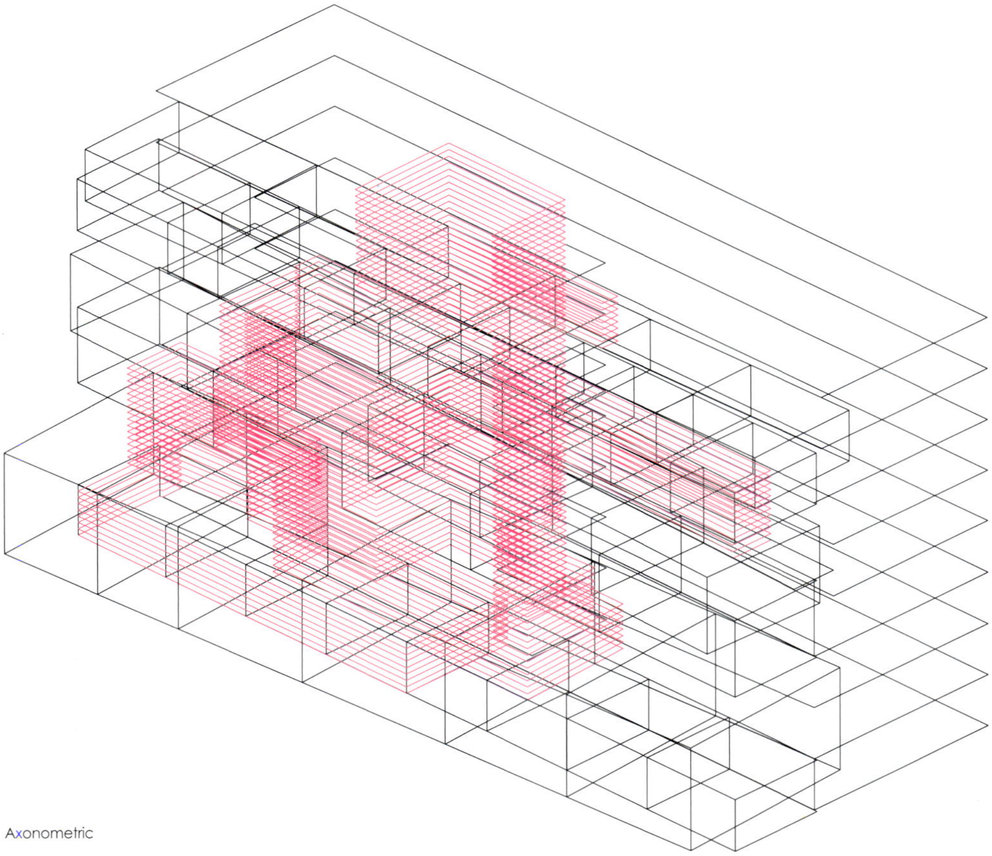

Axonometric

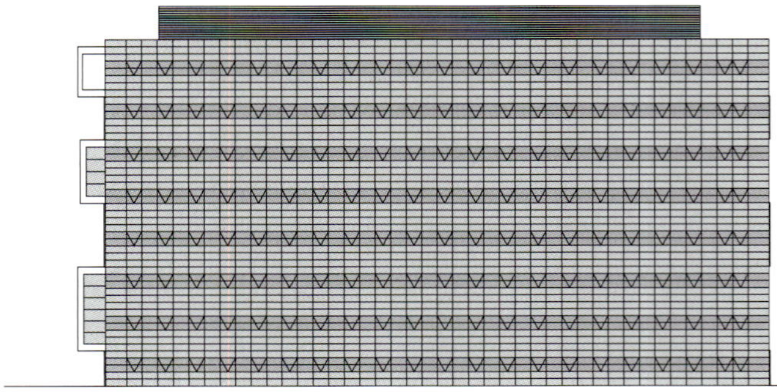

Elevation North

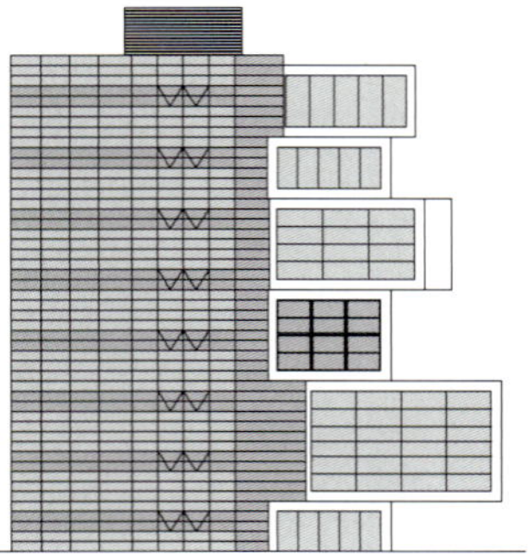

Elevation West

Chaoyang Park Plaza, Beijing, China

MAD Architects

© Iwan Baan

© Iwan Baan

© Iwan Baan

© Iwan Baan

© Iwan Baan

© Iwan Baan

© Iwan Baan

© Iwan Baan

© Iwan Baan

© Iwan Baan

Community Centre Aussersihl Part2, Zurich, Switzerland

EM2N

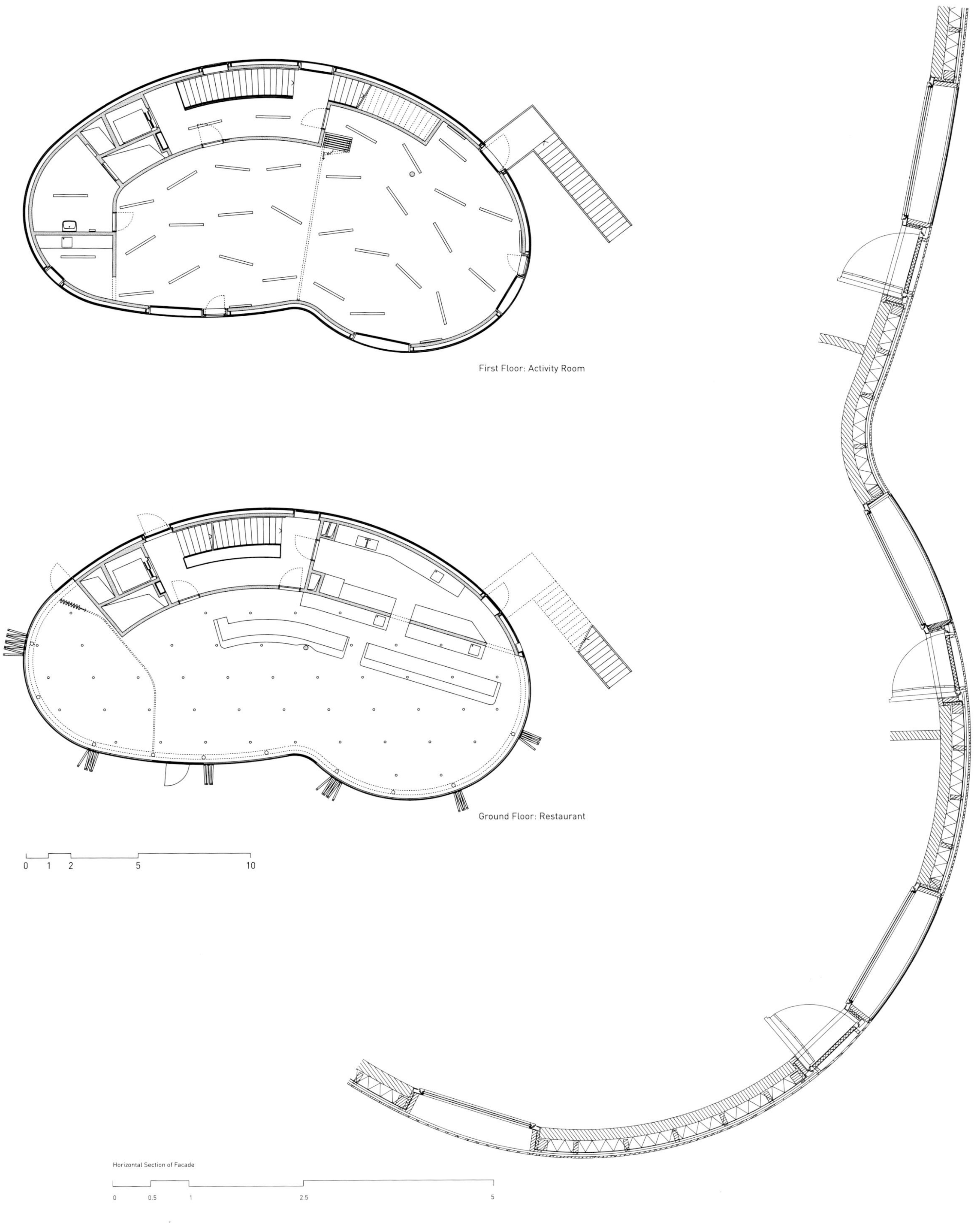
First Floor: Activity Room
Ground Floor: Restaurant
0
1
2
5
10
Horizontal Section of Facade
0
0.5
1
2.5
5

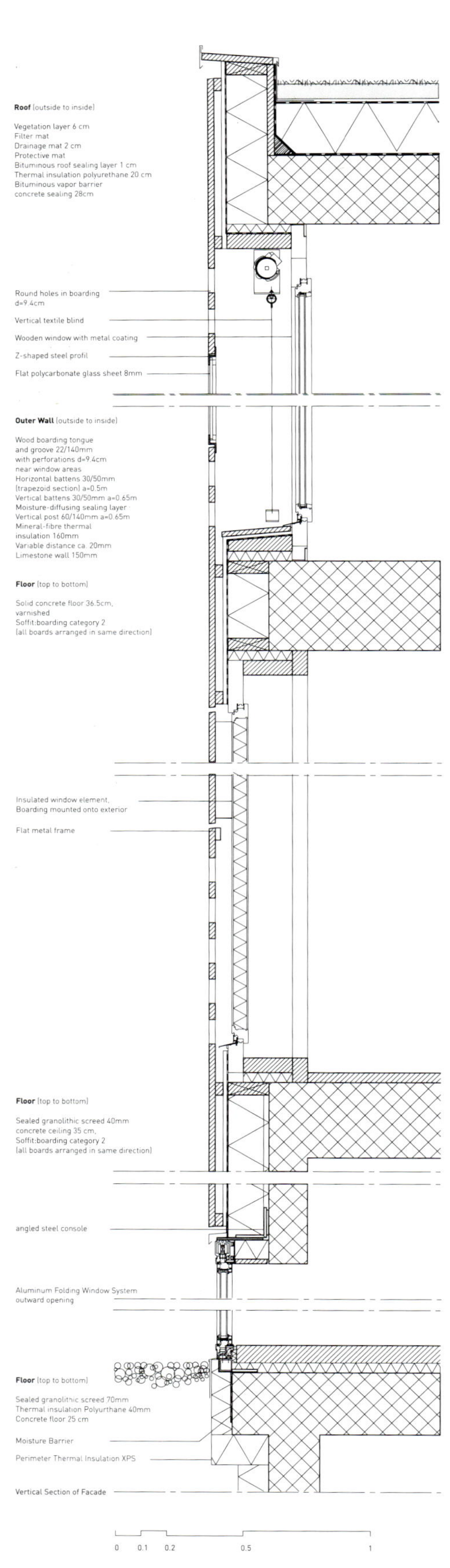

Roof (outside to inside)
Vegetation layer 6 cm
Filter mat
Drainage mat 2 cm
Protective mat
Bituminous roof sealing layer 1 cm
Thermal insulation polyurethane 20 cm
Bituminous vapor barrier
concrete sealing 28cm
Round holes in boarding
d=9.4cm
Vertical textile blind
Wooden window with metal coating
Z-shaped steel profil
Flat polycarbonate glass sheet 8mm
Outer Wall (outside to inside)
Wood boarding tongue
and groove 22/140mm
with perforations d=9.4cm
near window areas
Horizontal battens 30/50mm
(trapezoid section) a=0.5m
Vertical battens 30/50mm a=0.65m
Moisture-diffusing sealing layer
Vertical post 60/140mm a=0.65m
Mineral-fibre thermal
insulation 160mm
Variable distance ca. 20mm
Limestone wall 150mm
Floor (top to bottom)
Solid concrete floor 36.5cm,
varnished
Soffit:boarding category 2
(all boards arranged in same direction)
Insulated window element,
Boarding mounted onto exterior
Flat metal frame
Floor (top to bottom)
Sealed granolithic screed 40mm
concrete ceiling 35 cm,
Soffit:boarding category 2
(all boards arranged in same direction)
angled steel console
Aluminum Folding Window System
outward opening
Floor (top to bottom)
Sealed granolithic screed 70mm
Thermal insulation Polyurthane 40mm
Concrete floor 25 cm
Moisture Barrier
Perimeter Thermal Insulation XPS
Vertical Section of Facade
0
0.1
0.2
0.5
1

Crystal Houses, Amsterdam, Netherlands
MVRDV

© Daria Scagliola & Stijn Brakkee

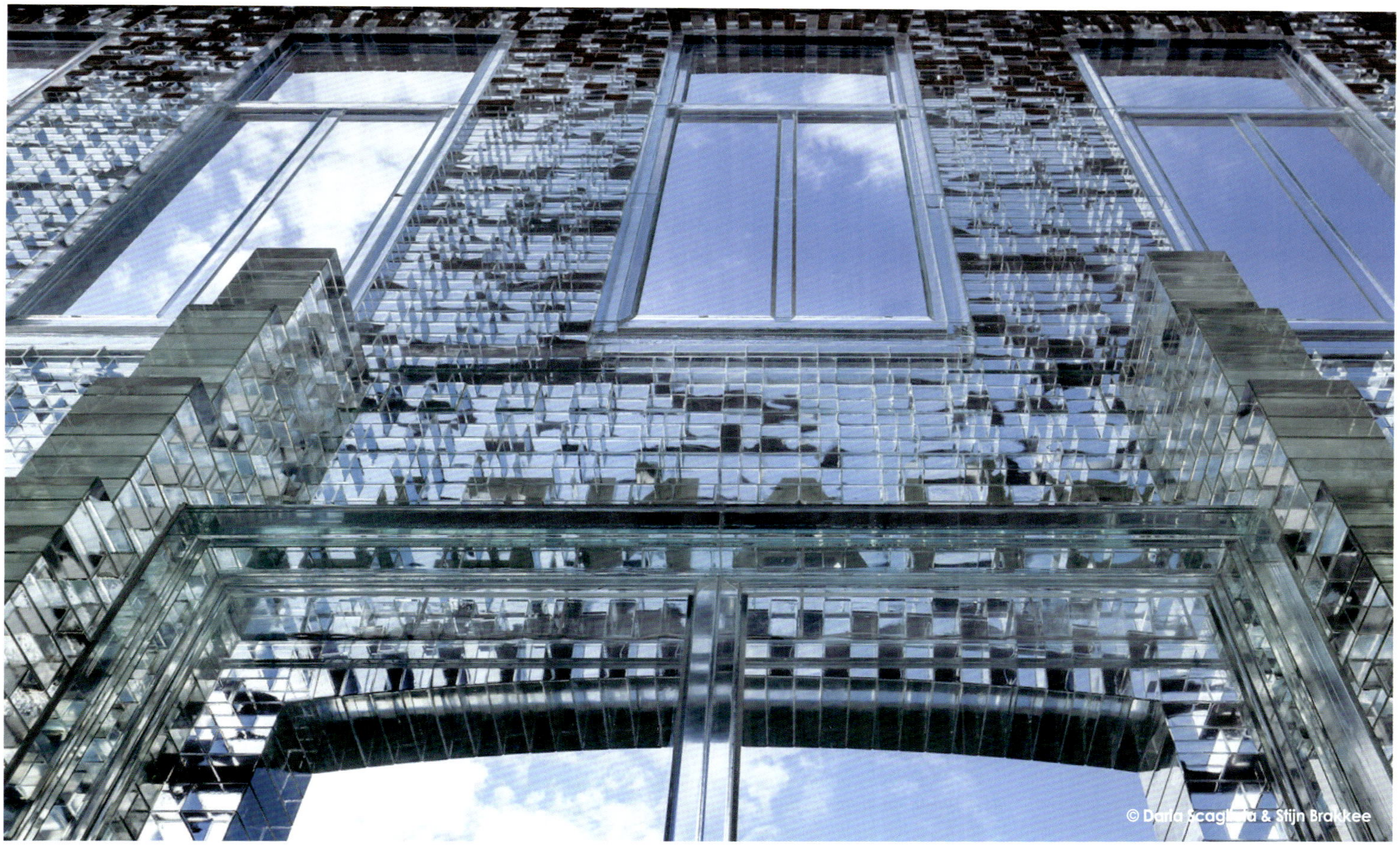
© Daria Scagliola & Stijn Brakkee

1.
© Poesia

2.
© MVRDV

3.
© MVRDV

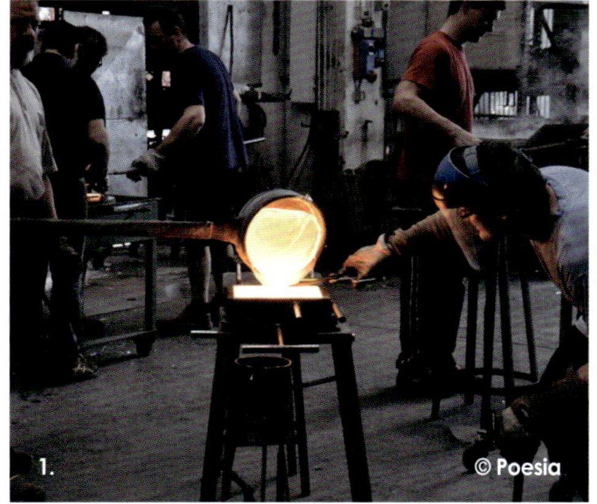

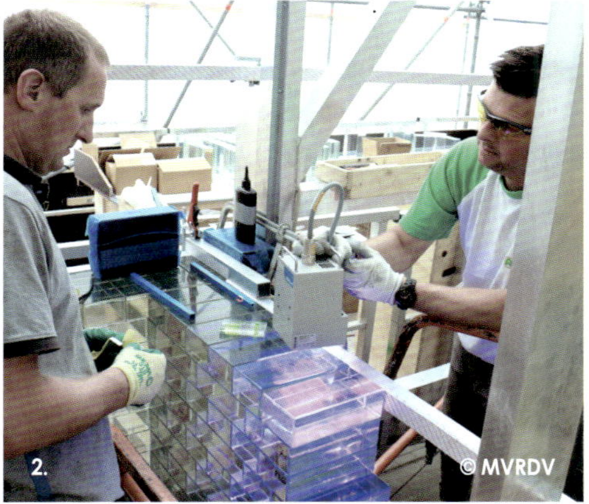

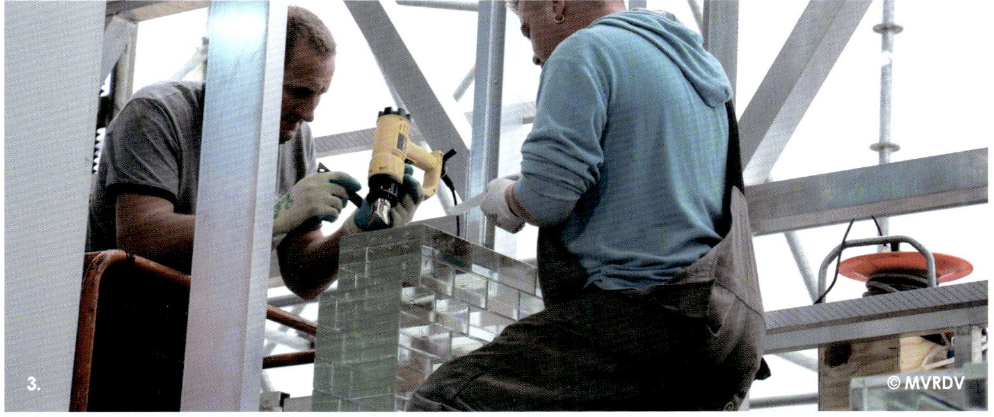

Construction Process

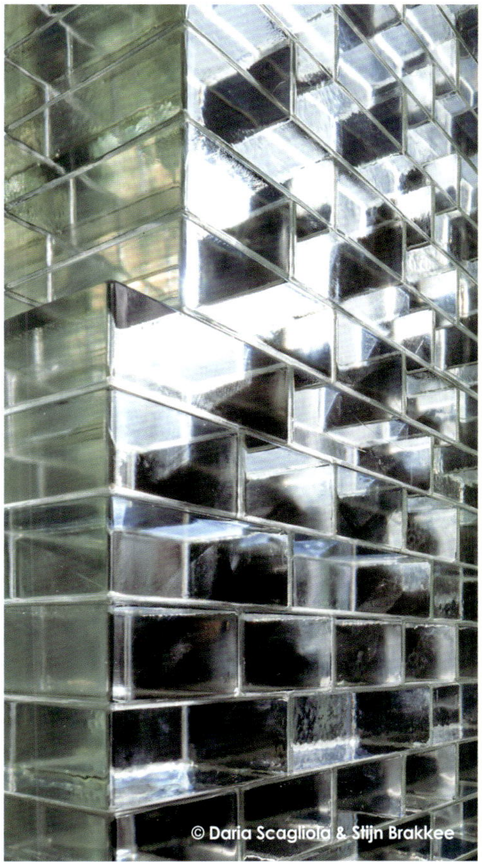
© Daria Scagliola & Stijn Brakkee

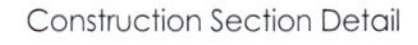

Construction Section Detail

© Daria Scagliola & Stijn Brakkee

MVRDV
CHANEL

Galleria Centercity, Cheonan, South Korea

UNStudio

Day View
Champagne silver
Back layer
Champagne Silver
Front layer
Champagne Silver
Night View in different colors
Back layer
White matte background
with artificial lighting
emphasizing wave area
Back layer
White matte background
with artificial lighting
emphasizing wave area
N
Different visualization using the colored lighting

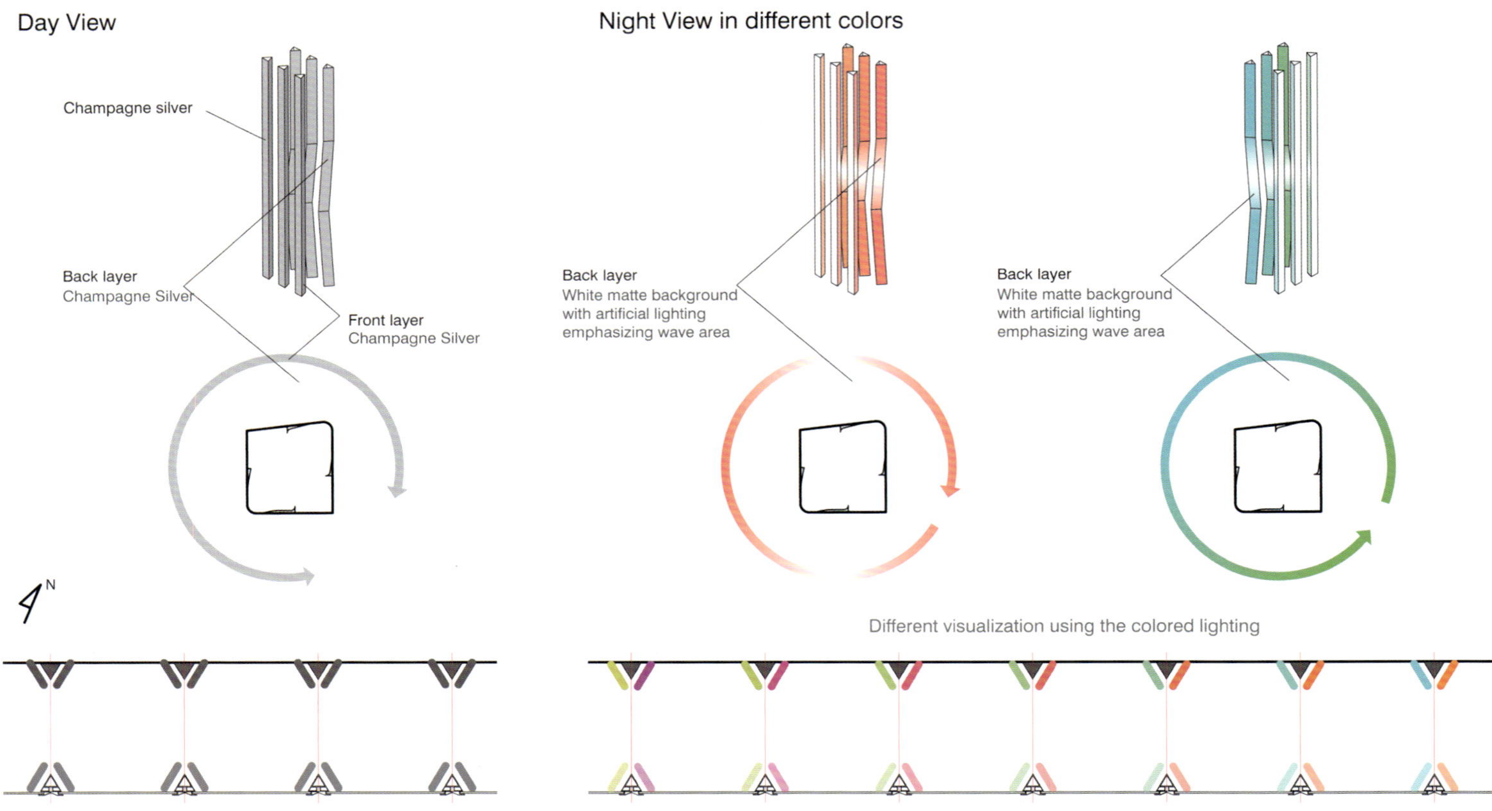

Color Scheme

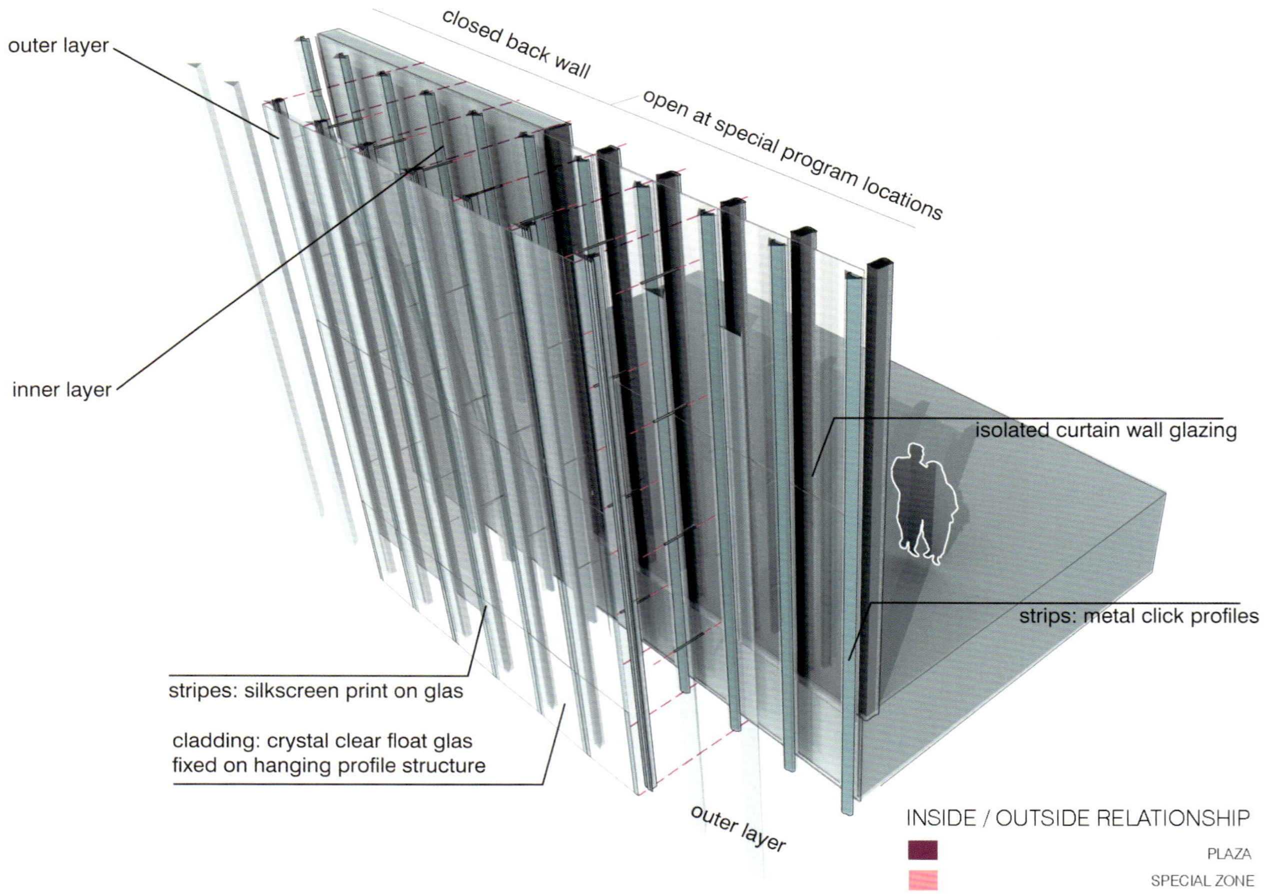

outer layer
closed back wall
open at special program locations
inner layer
isolated curtain wall glazing
strips: metal click profiles
stripes: silkscreen print on glas
cladding: crystal clear float glas
fixed on hanging profile structure
outer layer
INSIDE / OUTSIDE RELATIONSHIP
PLAZA
SPECIAL ZONE

Facade Concept

© Christian Richters

© Christian Richters

© Christian Richters

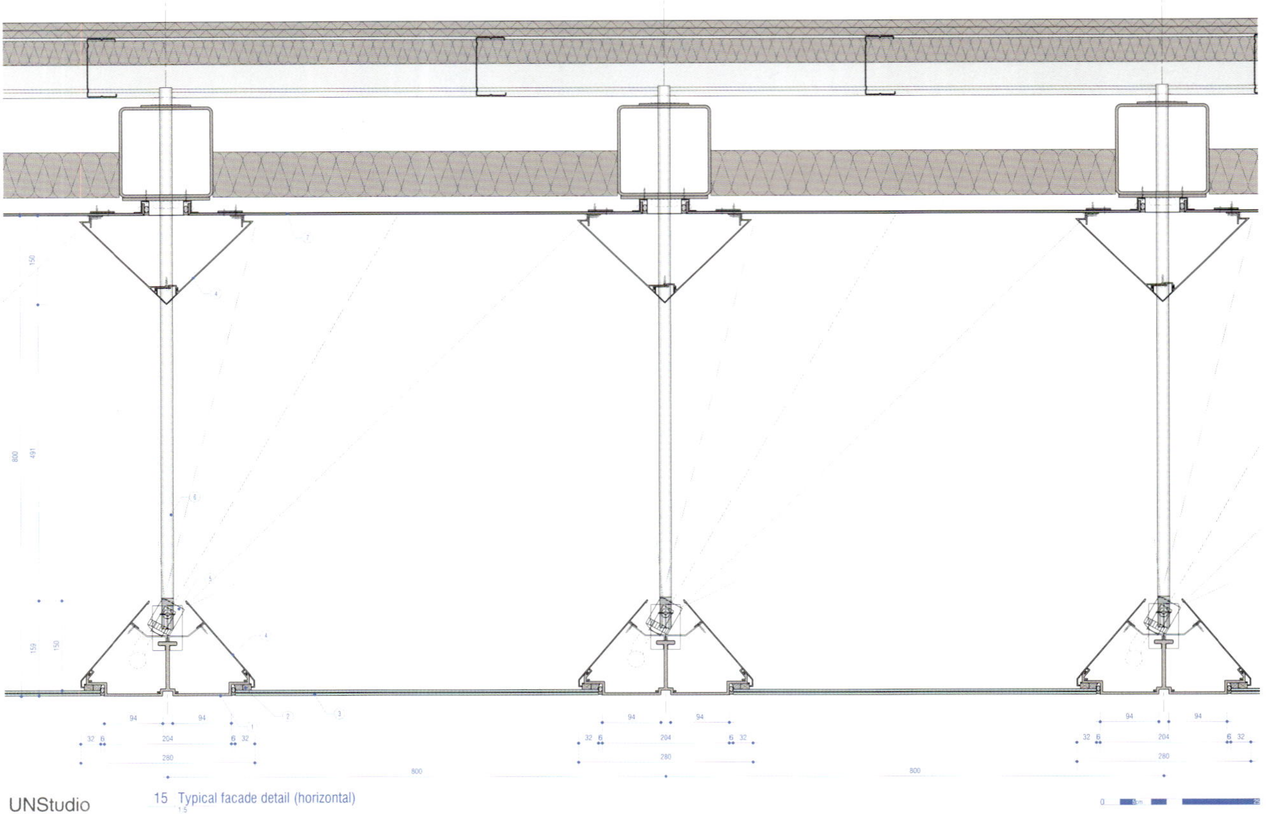

Facade Details

© Christian Richters

Facade Details

© Christian Richters

© Christian Richters

© Christian Richters

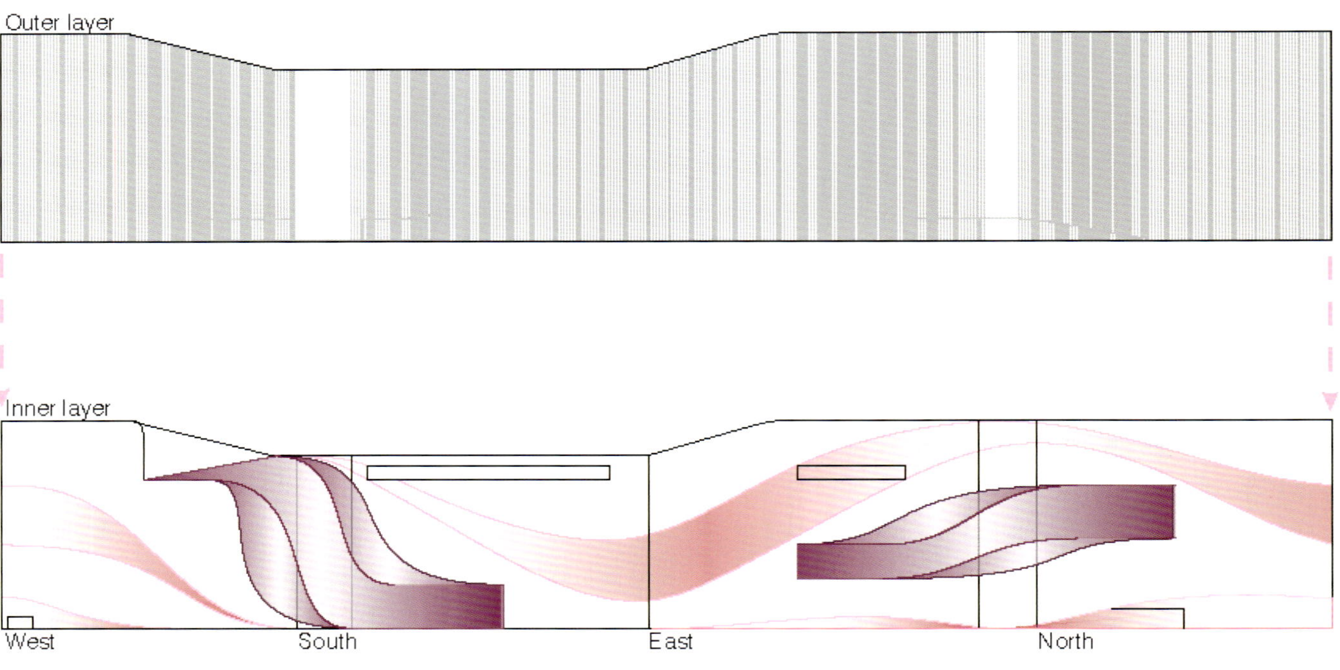

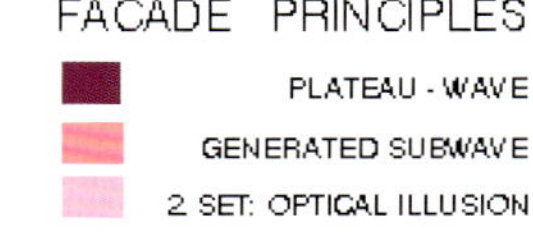

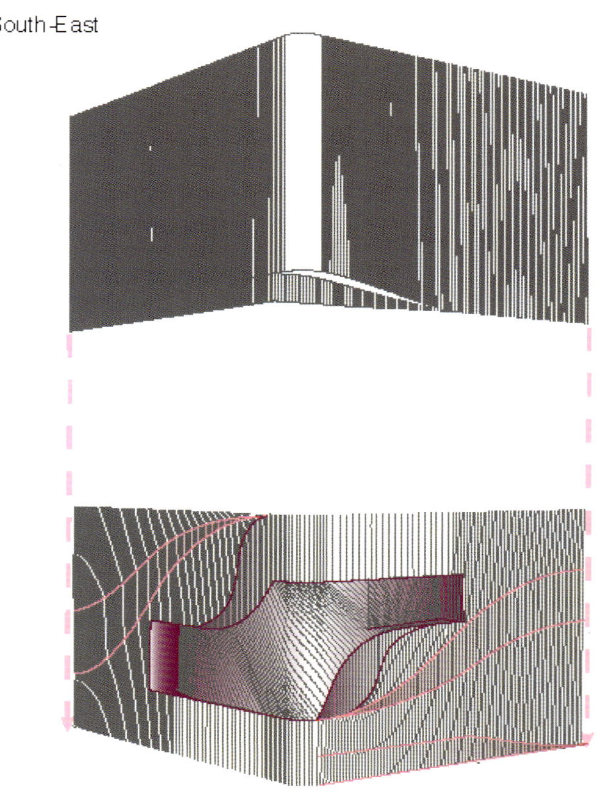

Facade Concept2

Patrizia Headquarters, Augsburg, Germany

kadawittfeldarchitektur

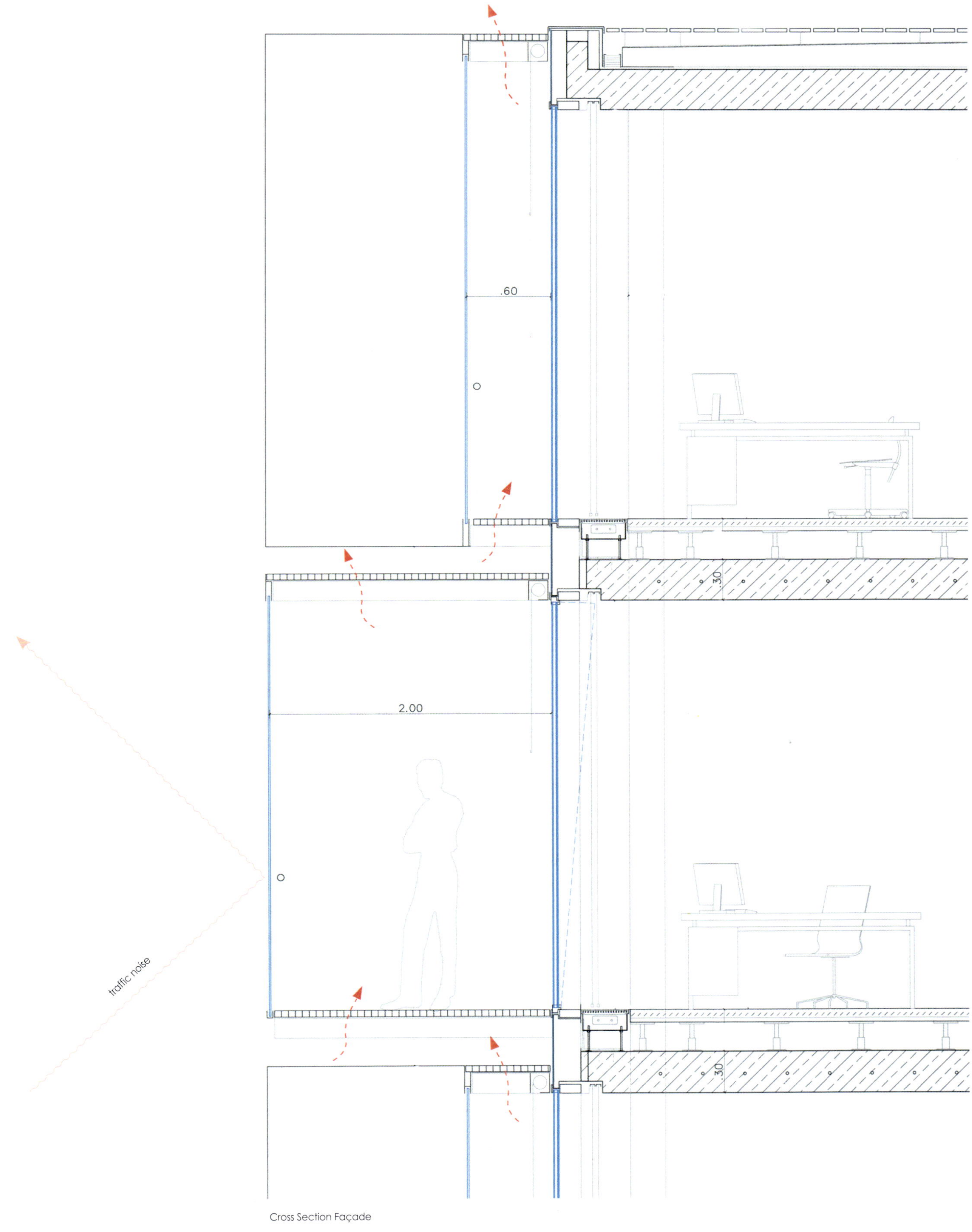

Cross Section Façade

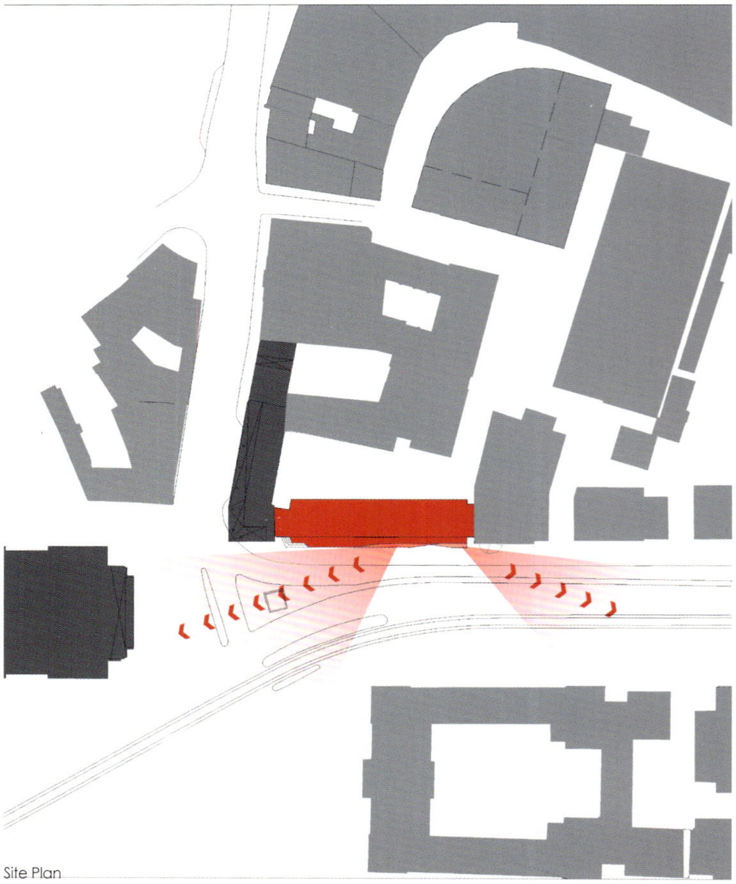

Second Floor

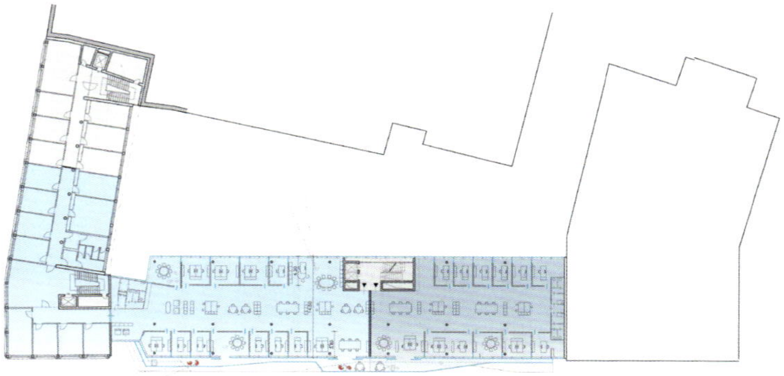

Third Floor

Site Plan

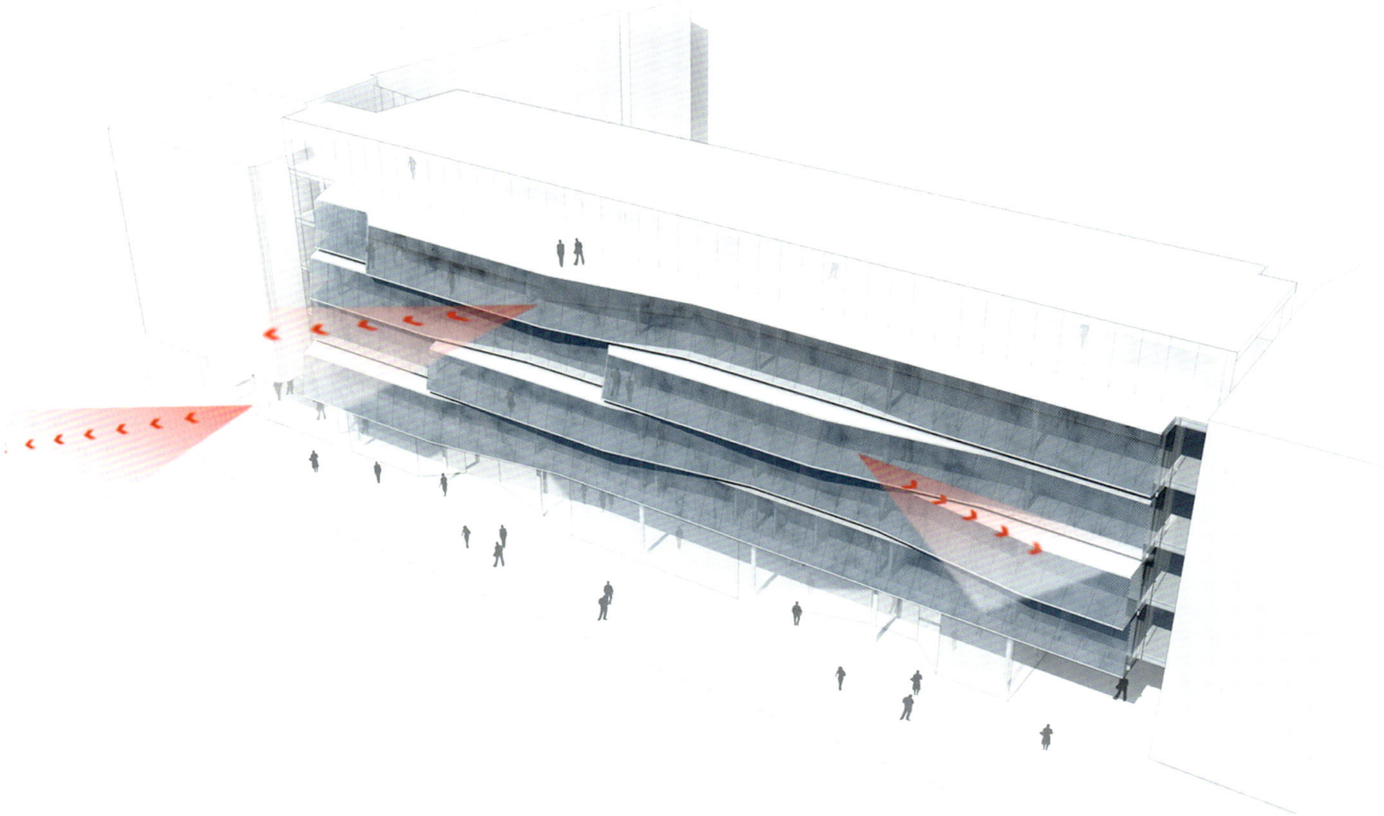

Rodin Gallery, Seoul, Korea

Kevin Kennon Architects

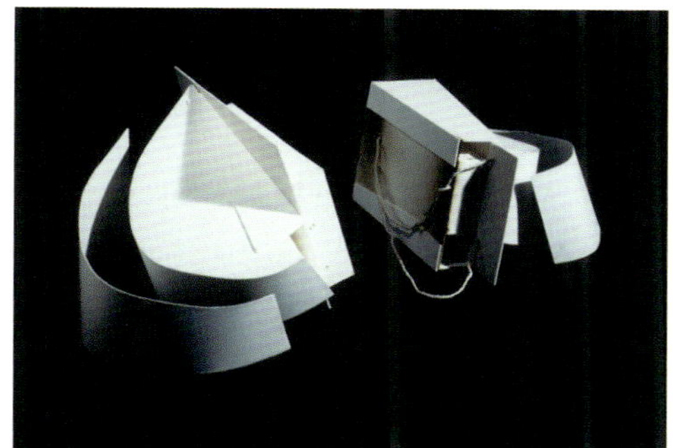

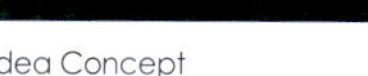

Idea Concept

Study Modeling

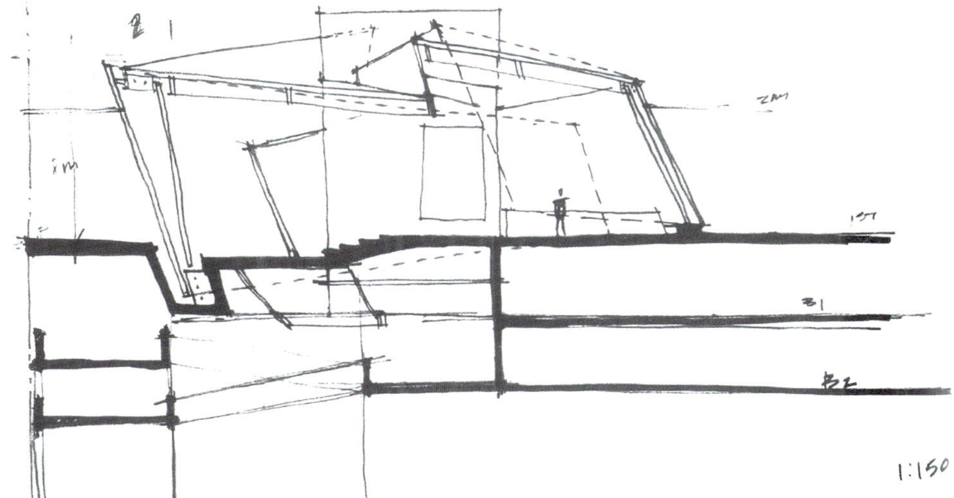

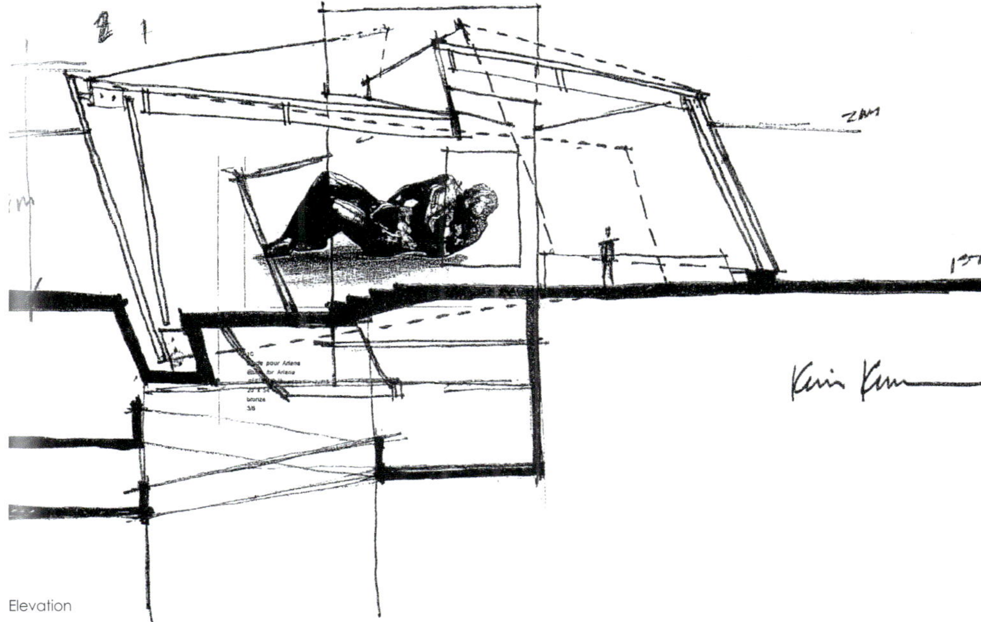

Elevation

Taghaus, bei Duesseldorf, Germany

Cheungvogl

Tiffin Bay, Kumpur, Malaysia

design spirits co., ltd.

UW:ads1a, Cologne, Germany
Bernd Kniess Architects Urban Planners

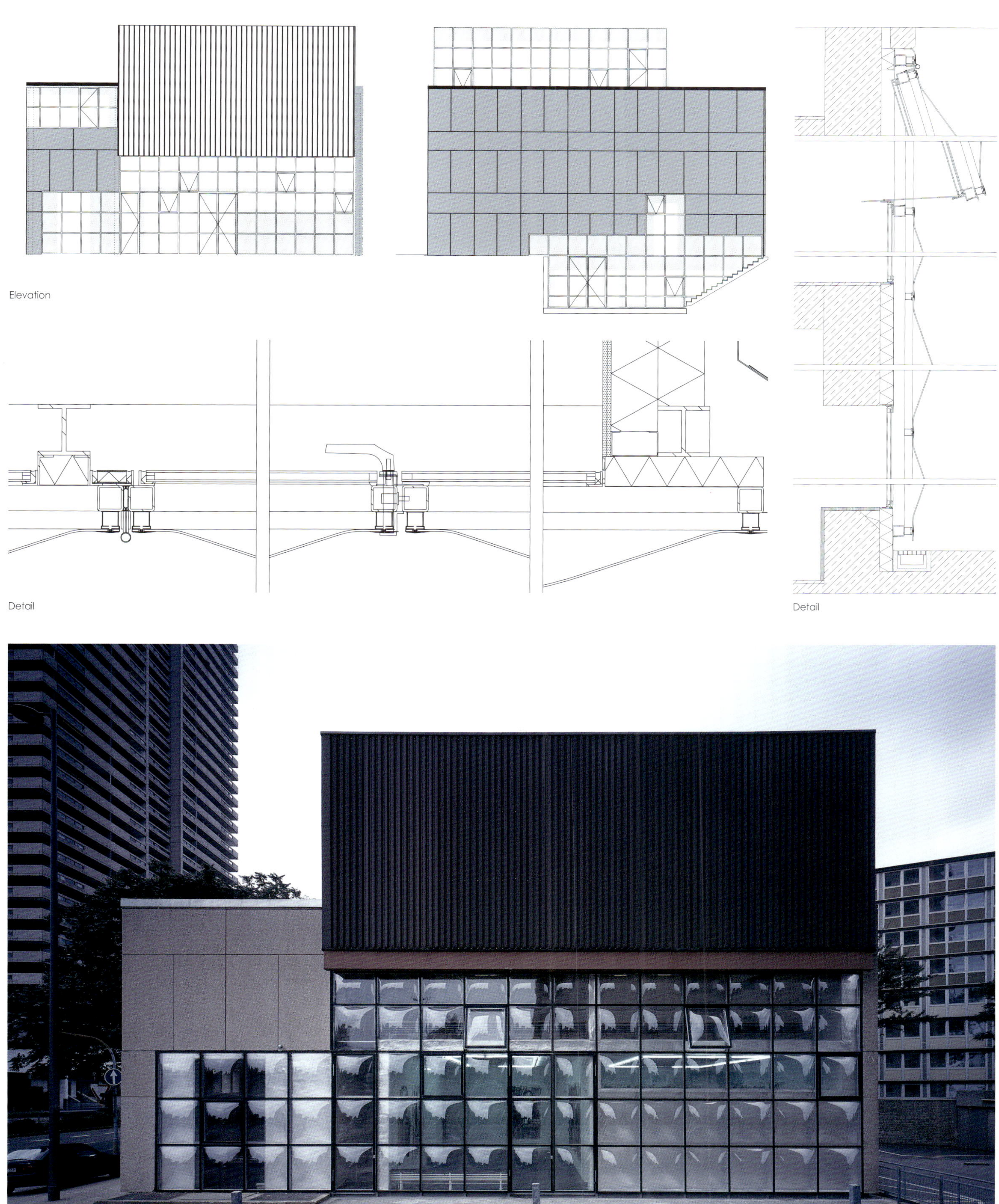
Elevation

Detail

Detail

Vip - Center Schiphol Airport, Schiphol, the Netherlands

Concrete Architectural Associates

© Ewout Huibers

Warehouse 17C, Madrid, Spain

Arturo Franco, Fabrice van Teslaar

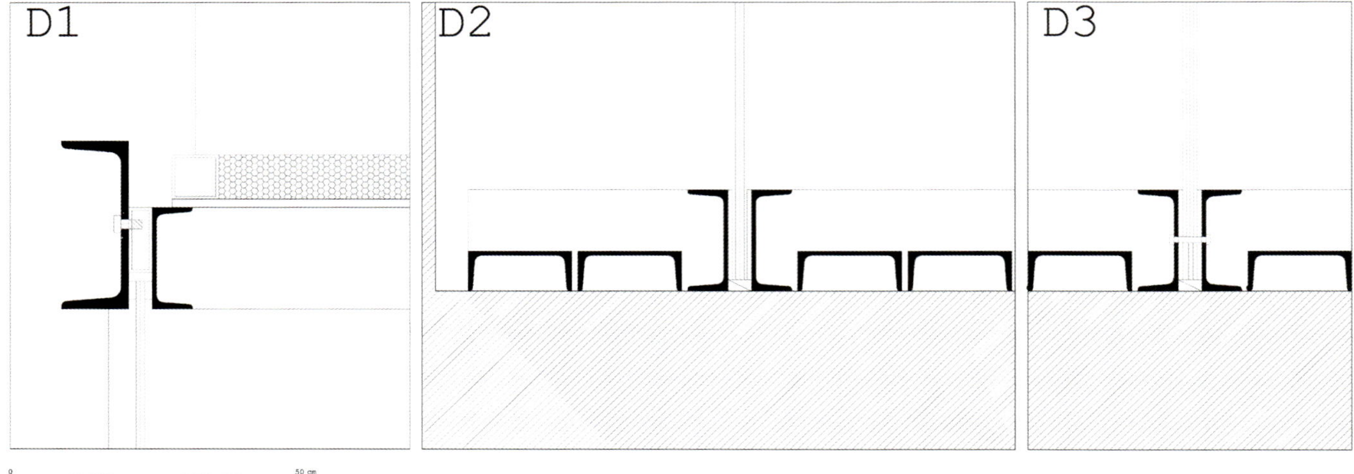

Glass Detail

Construction Process

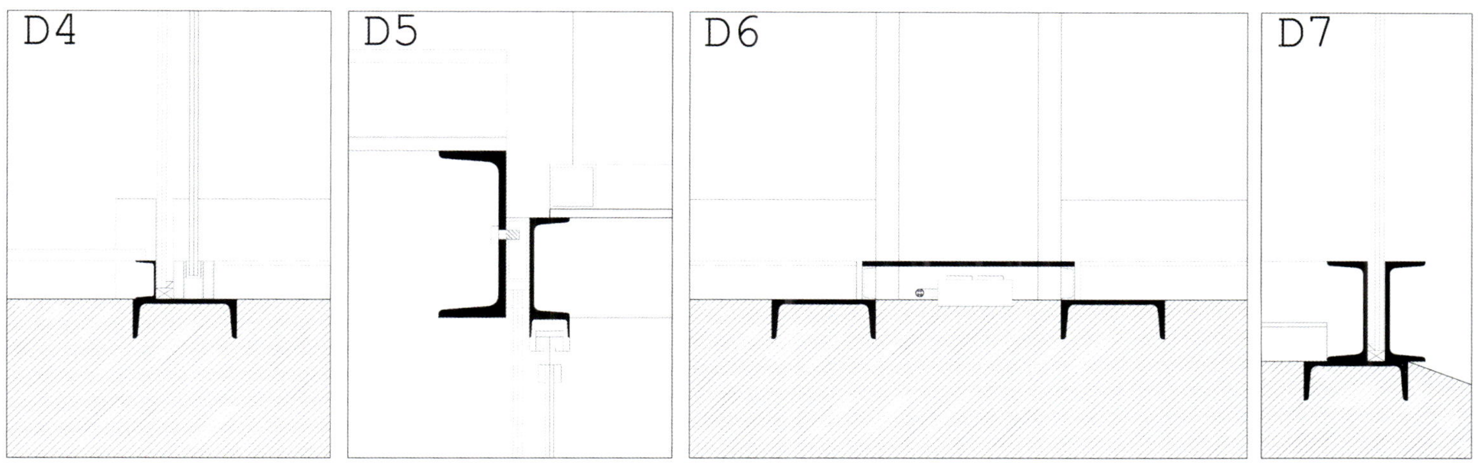

3.

6.

7.

Corniche Tower,

marco hemmerling architecture design

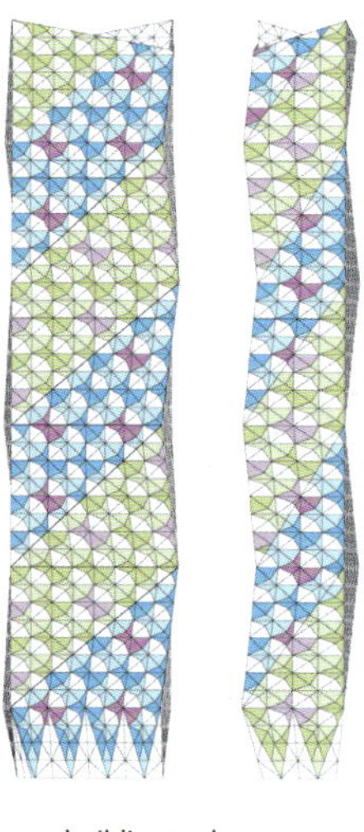
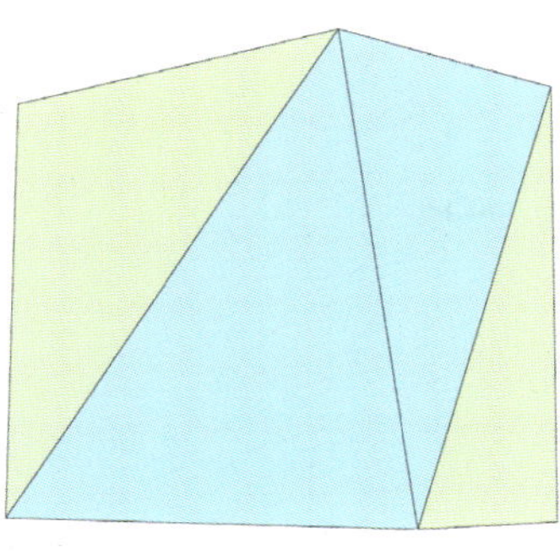
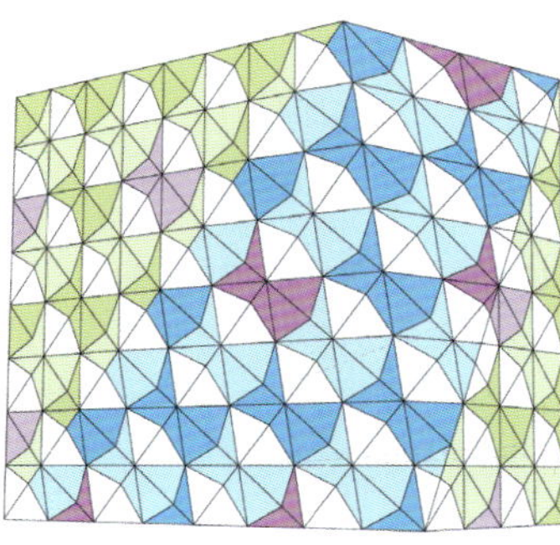
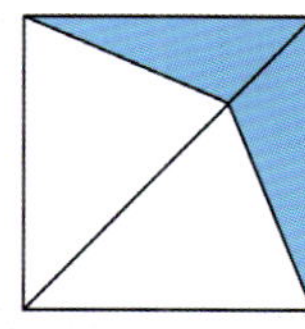

building scale

total facade area
~16965 sqm

modul scale1

8 storeys, 2 main spirals

modul facade area
~3240 sqm = 100%

modul scale 2

- 8 storeys, 2 main spirals
- 6 different colored panels, highly reflective, low degree of transparency
- 1 non-colored glass panel, transparent

glas panels (area)
green light ~564 sqm
green ~420sqm
blue light ~567 sqm = 66.3%
blue ~480sqm
red light ~133sqm
red ~ 130sqm

transparent glas ~1091sqm = 33.7% clear transparency

facade element scale

4 glas elements, in this case 2 blue and two transparent elements

Facade Study

Dolce & Gabbana Offices in via Broggi, Milan, Italy

Piuarch

First Floor

Top Floor

© Ruy Teixeira

© Alberto Piovano

© Alberto Piovano

Section

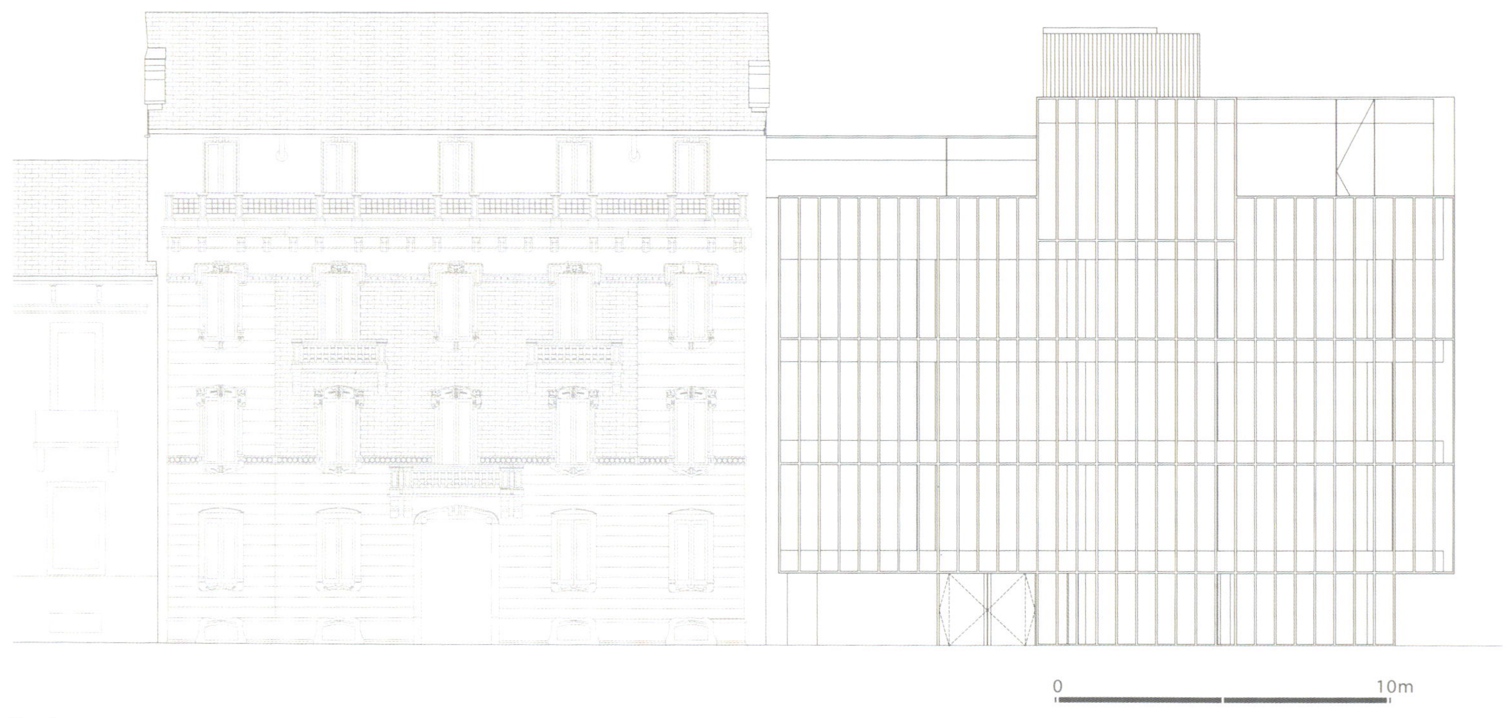

Elevation

© Ruy Teixeira

© Alberto Piovano

© Andrea Martiradonna

© Ruy Teixeira

Punto

Fratelli Wines Industrial Projects, Maharastra, India

Sunil Patil & Associates

© Hemant Patil

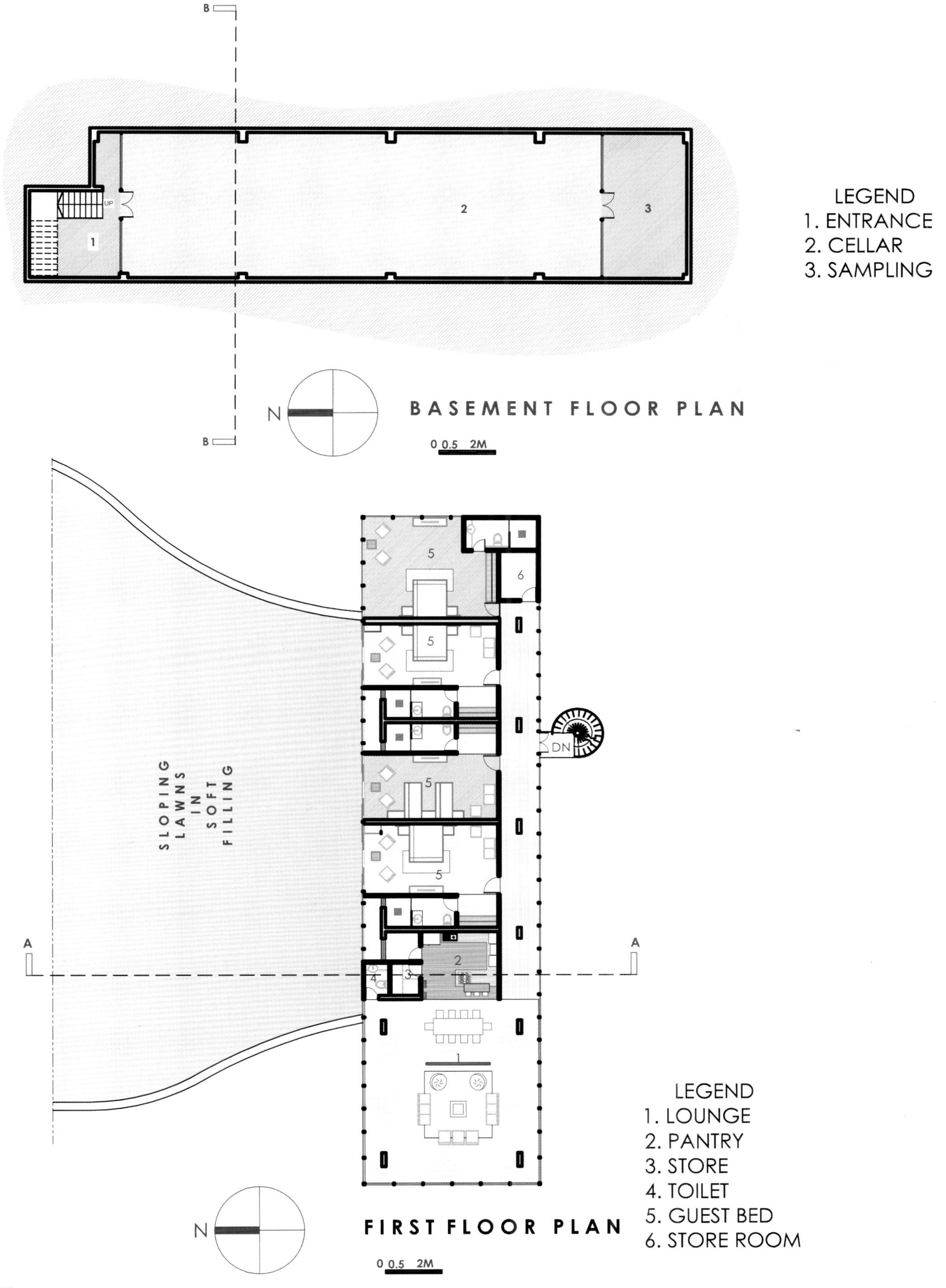

Administer Plan

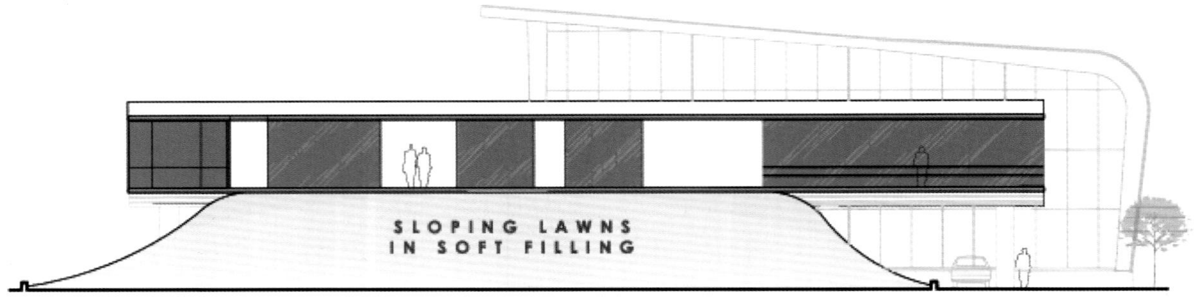

North Side Elevation

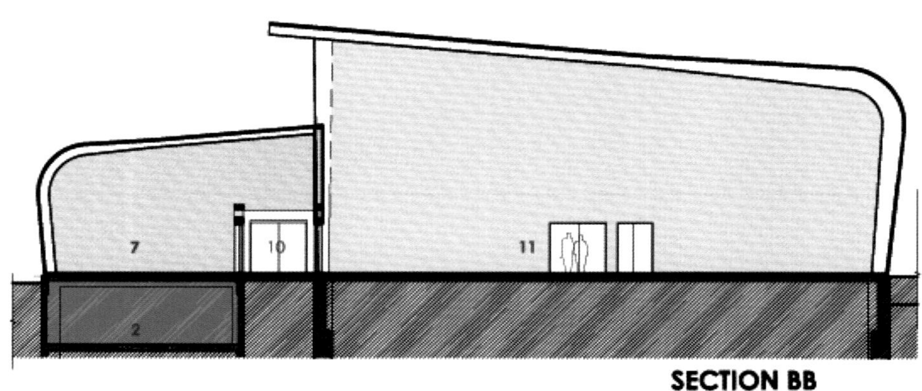

Section B-B

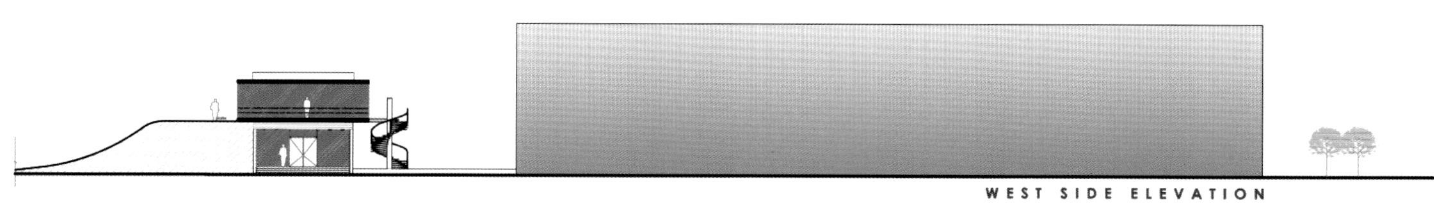

West Side Elevation

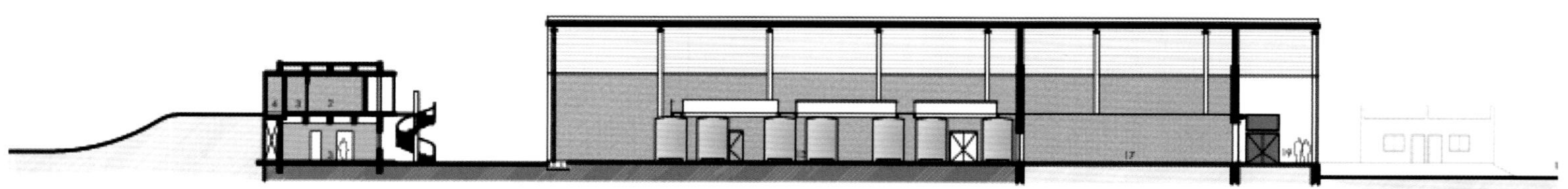

Elevation A-A

© Hemant Patil

© Hemant Patil

© Hemant Patil

© Hemant Patil

© Hemant Patil

© Hemant Patil

FRATELLI
FRATELLI

© Hemant Patil

GEN Design Studio office, Braga, Portugal

murmuro

© Pedro Nuno Pacheco

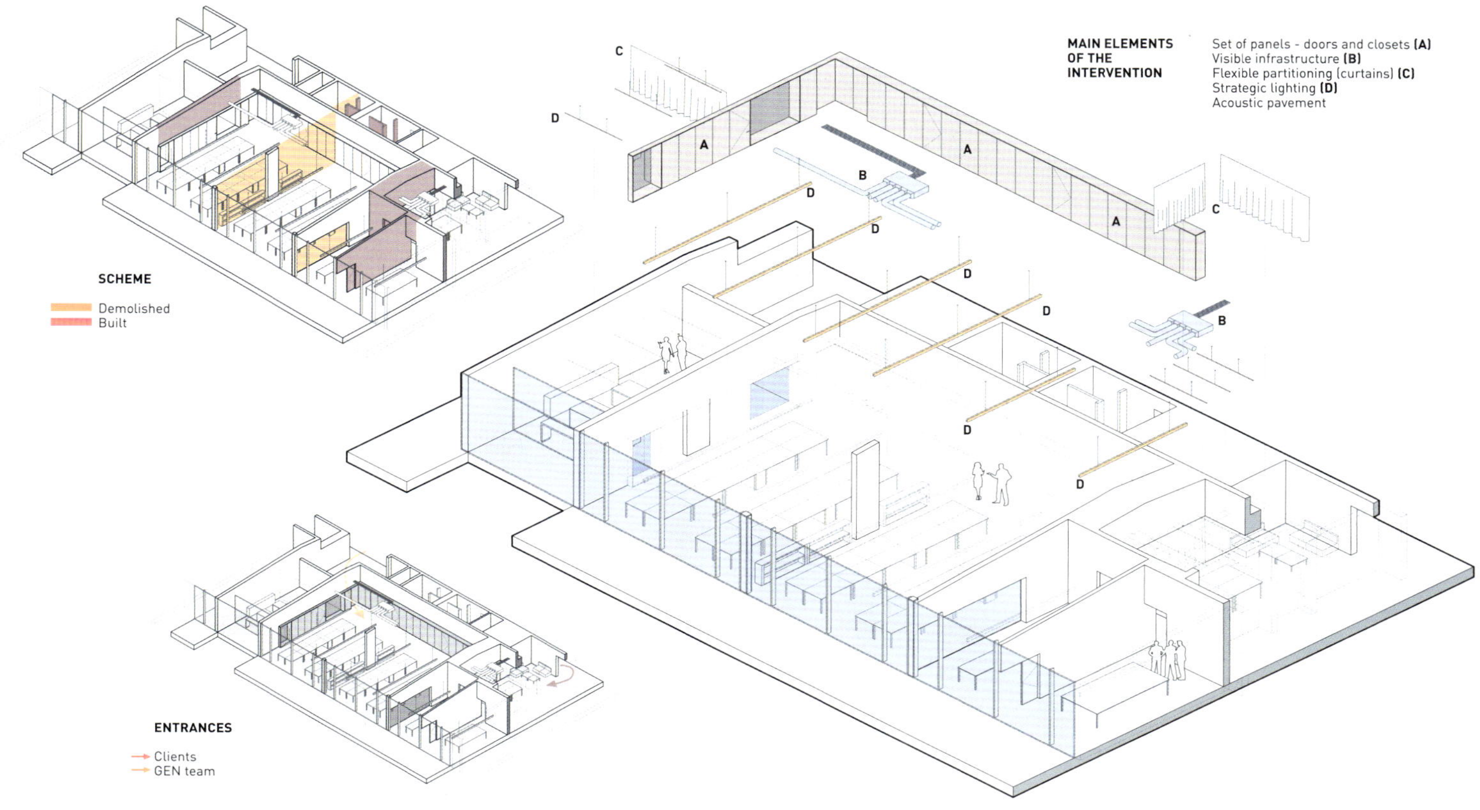

Axonometric

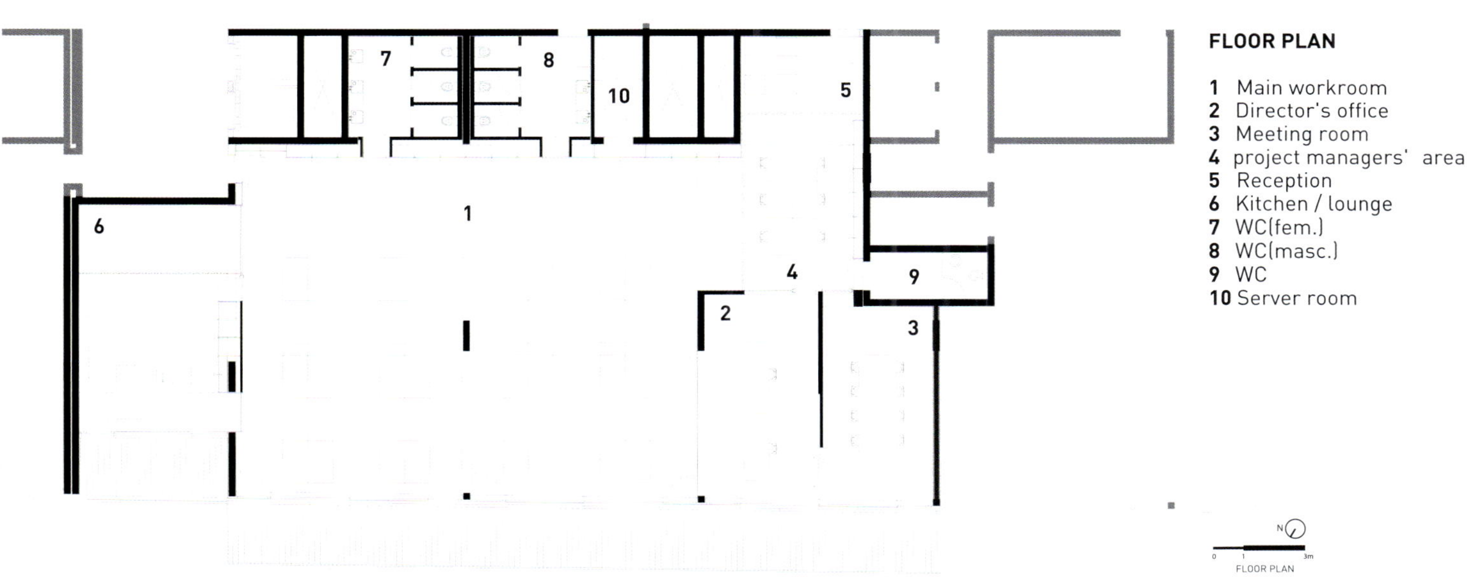

Floor Plan

© Pedro Nuno Pacheco

© Pedro Nuno Pacheco

© Pedro Nuno Pacheco

© Pedro Nuno Pacheco

High and Dry
BOARD

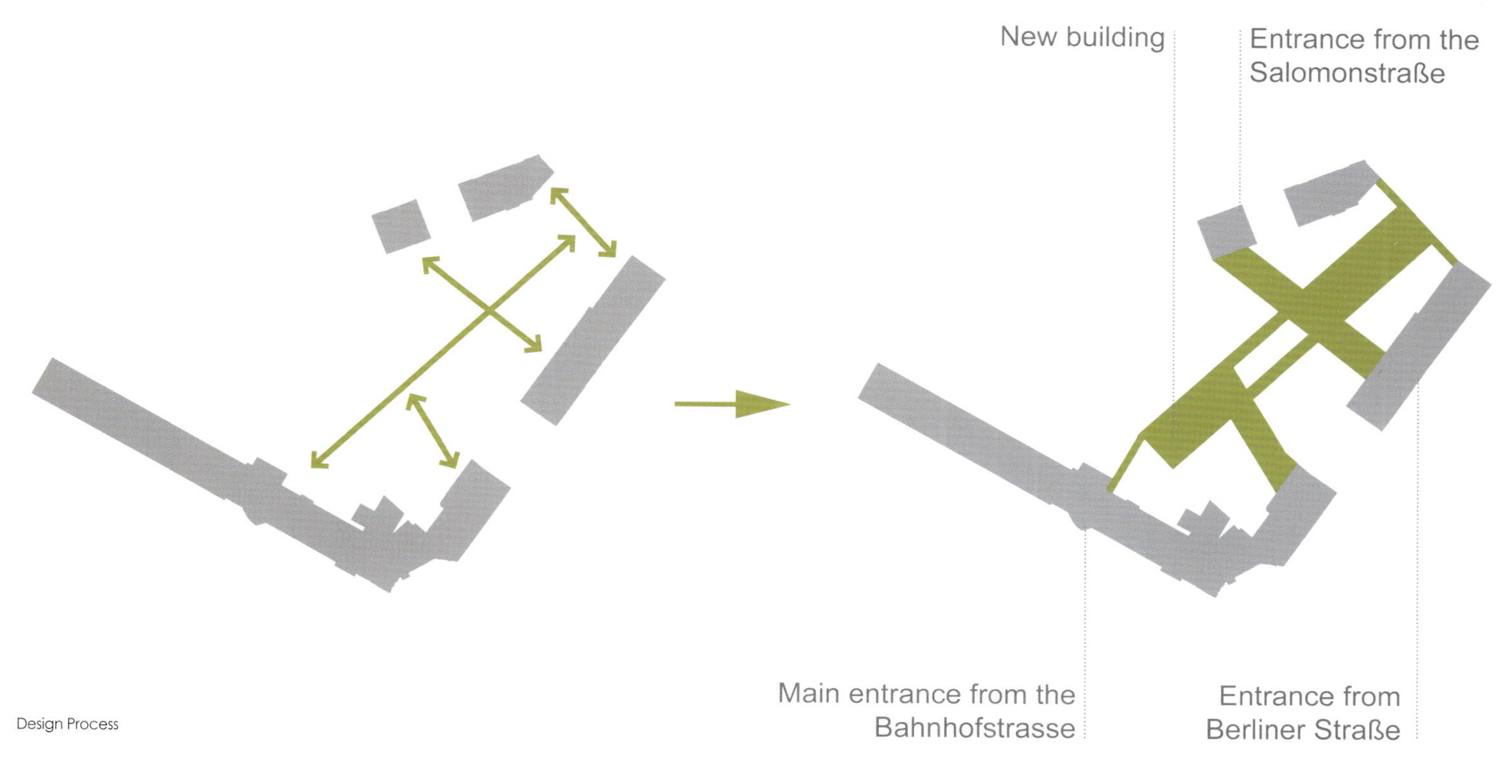

Design Process

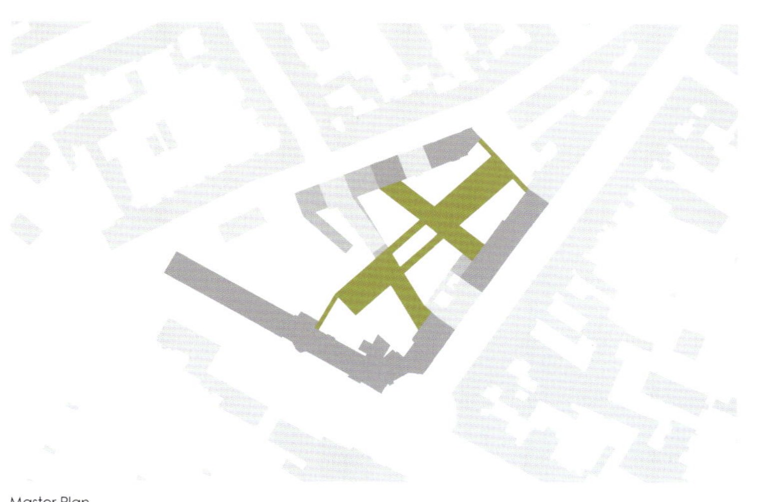

Master Plan

Elevation Design Process

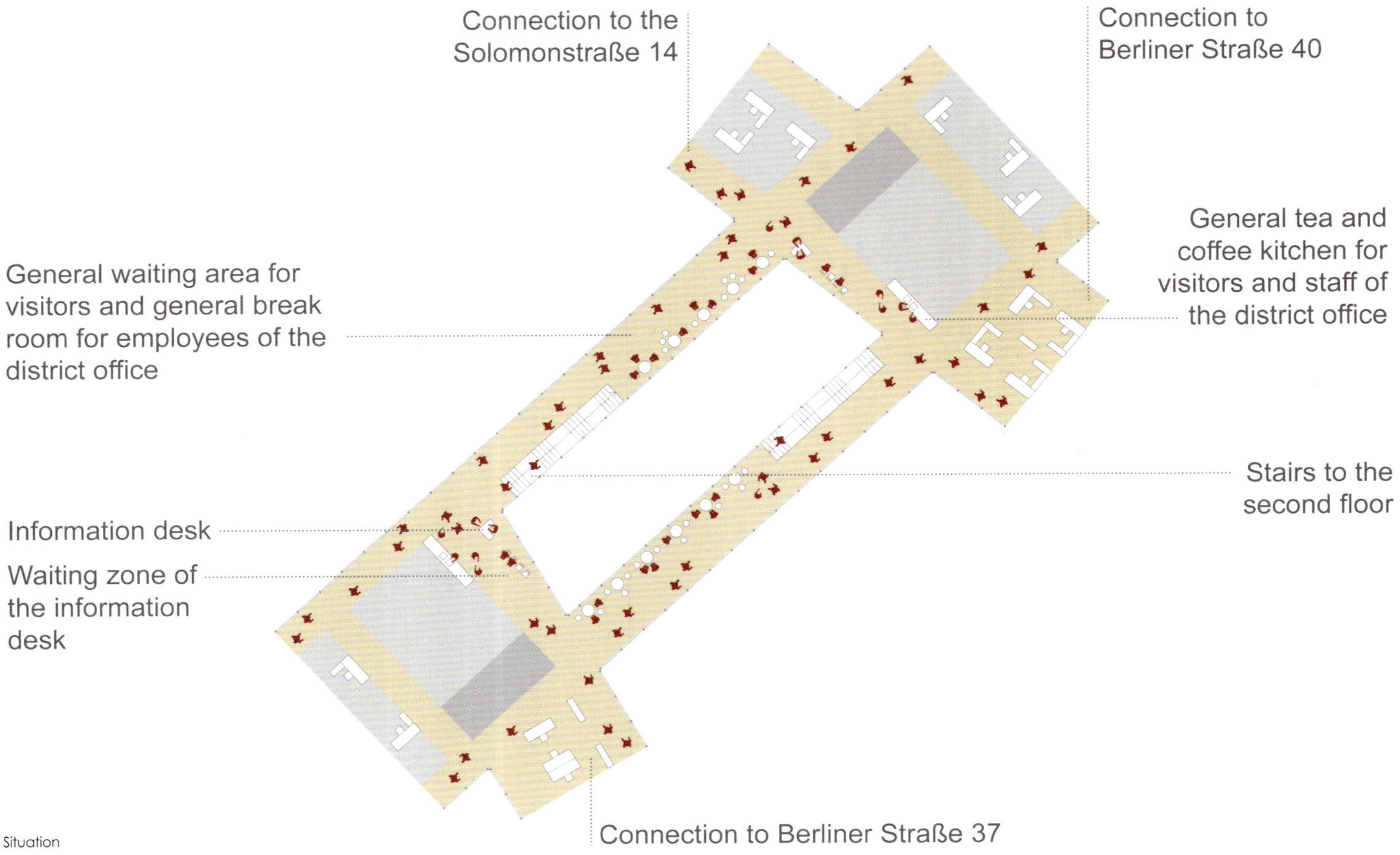

Connection to the
Solomonstraße 14
Connection to
Berliner Straße 40
General waiting area for
visitors and general break
room for employees of the
district office
General tea and
coffee kitchen for
visitors and staff of
the district office
Stairs to the
second floor
Information desk
Waiting zone of
the information
desk
Connection to Berliner Straße 37

Site Situation

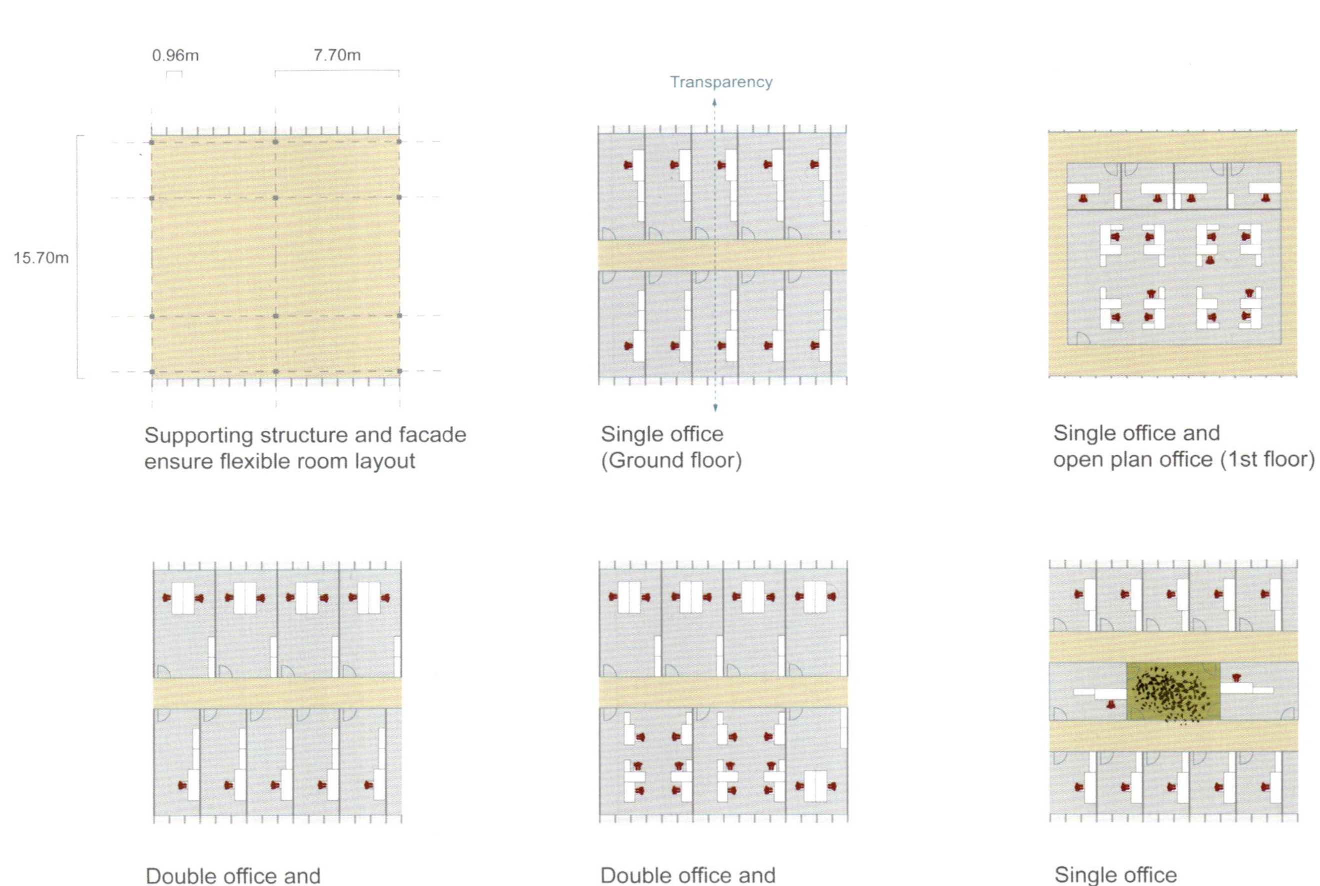

0.96m
7.70m
15.70m
Transparency
Supporting structure and facade
ensure flexible room layout
Single office
(Ground floor)
Single office and
open plan office (1st floor)
Double office and
single office (2nd floor)
Double office and
open plan office (2nd floor)
Single office
(3rd floor)

Site Situation2

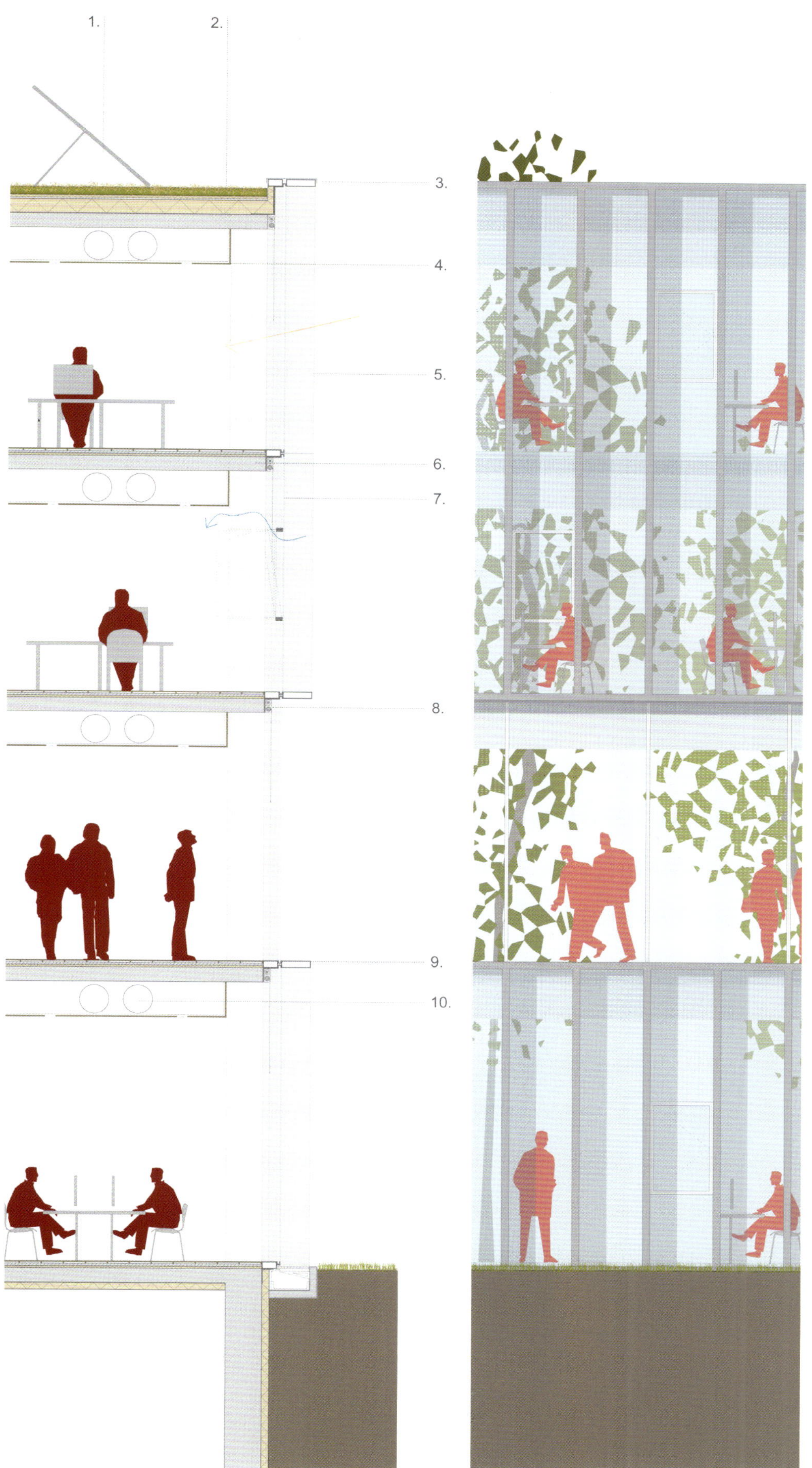

1. Photovoltaic system
2. Green roof
3. Aluminum sheet
4. Wooden acoustic ceiling
5. Brise soleil aluminum profile
6. Reinforced concrete floor
7. Triple glazing in aluminum frame
8. Sunshade made of stainless steel limbs
9. Wooden floor
10. Sustainable air conditioning with heat reco

Section Detail

Intecs Spa Headquarters, Roma, Italy

modostudio

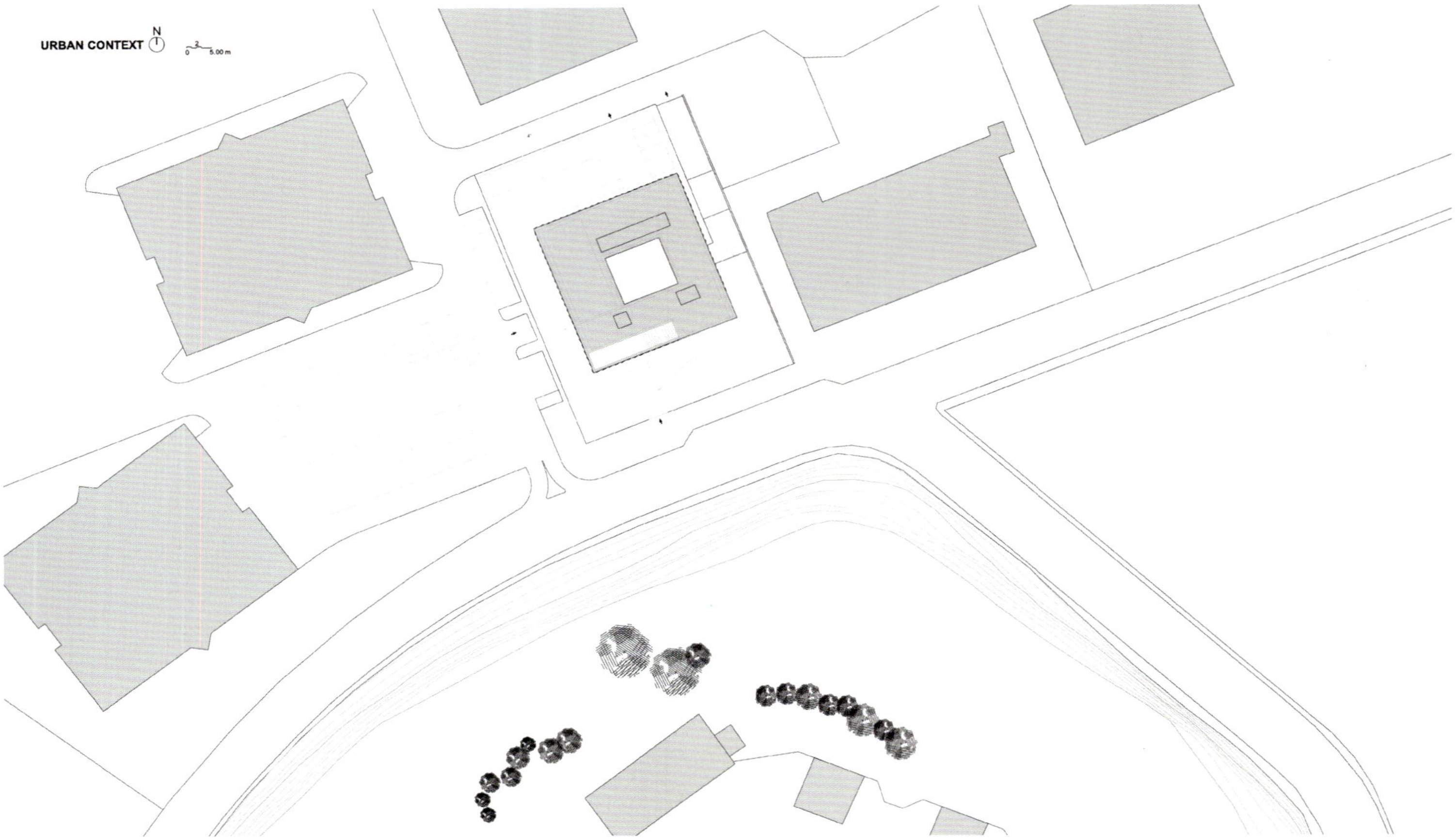

URBAN CONTEXT N 0 2 5.00 m

Facade Design Diagram

© SOLANGE SOUZA

South Elevation

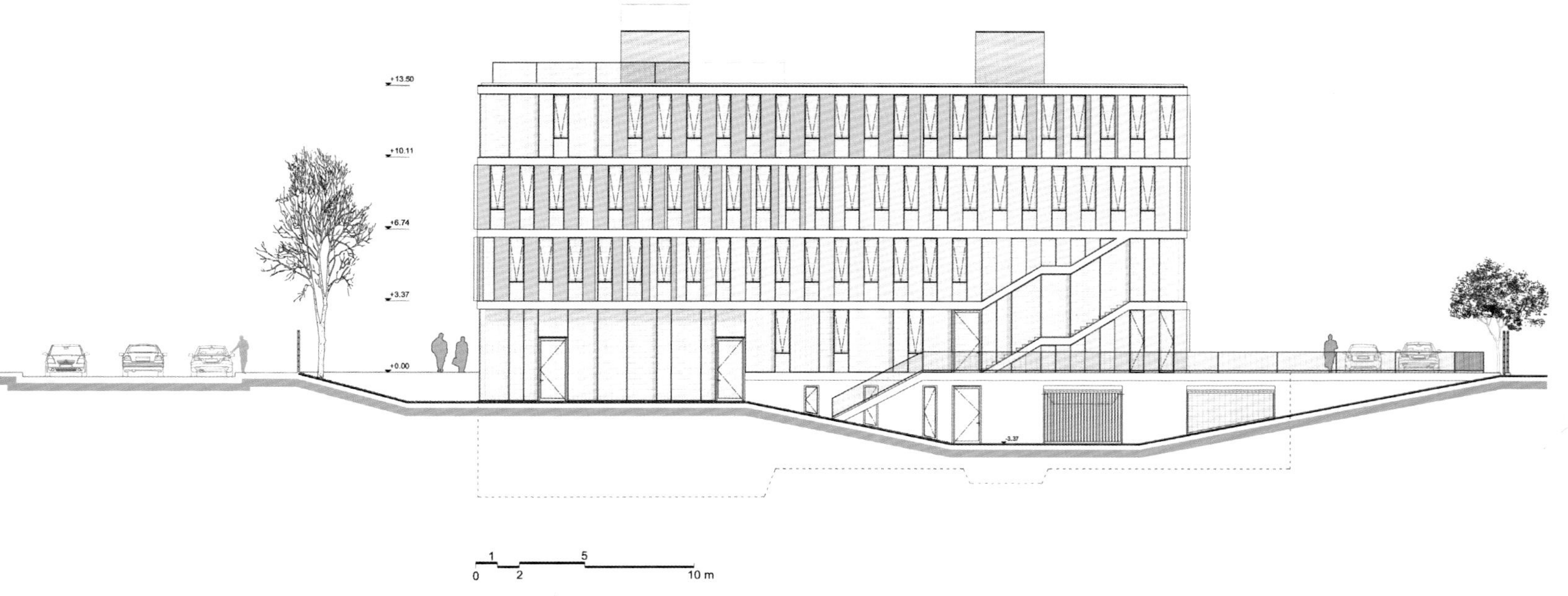

East Elevation

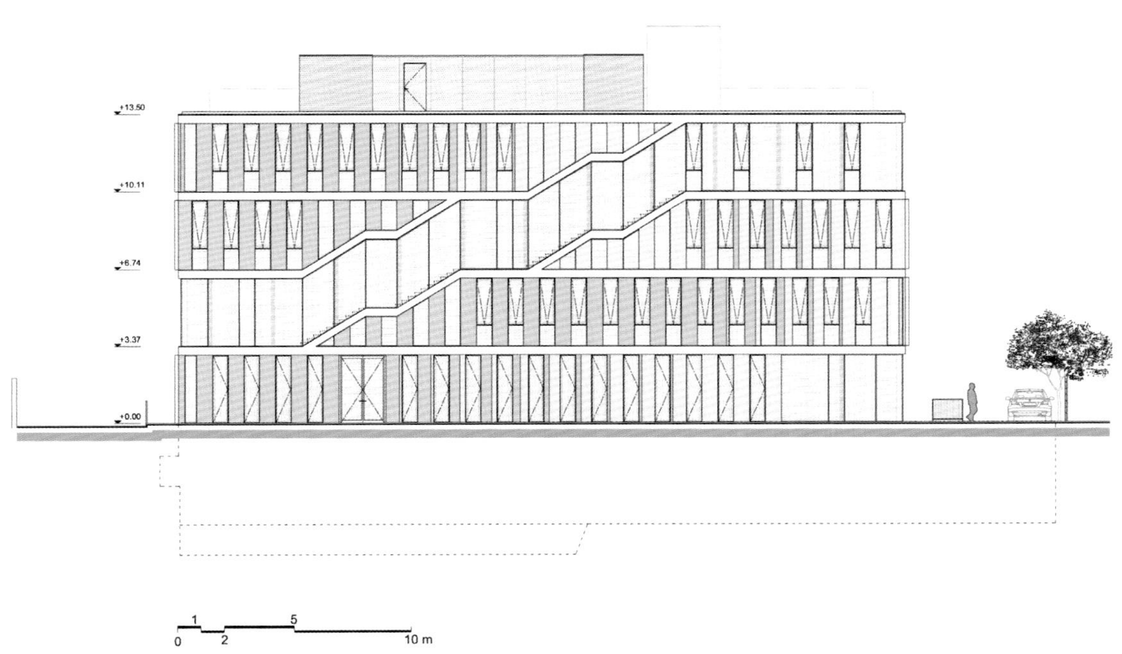

North Elevation

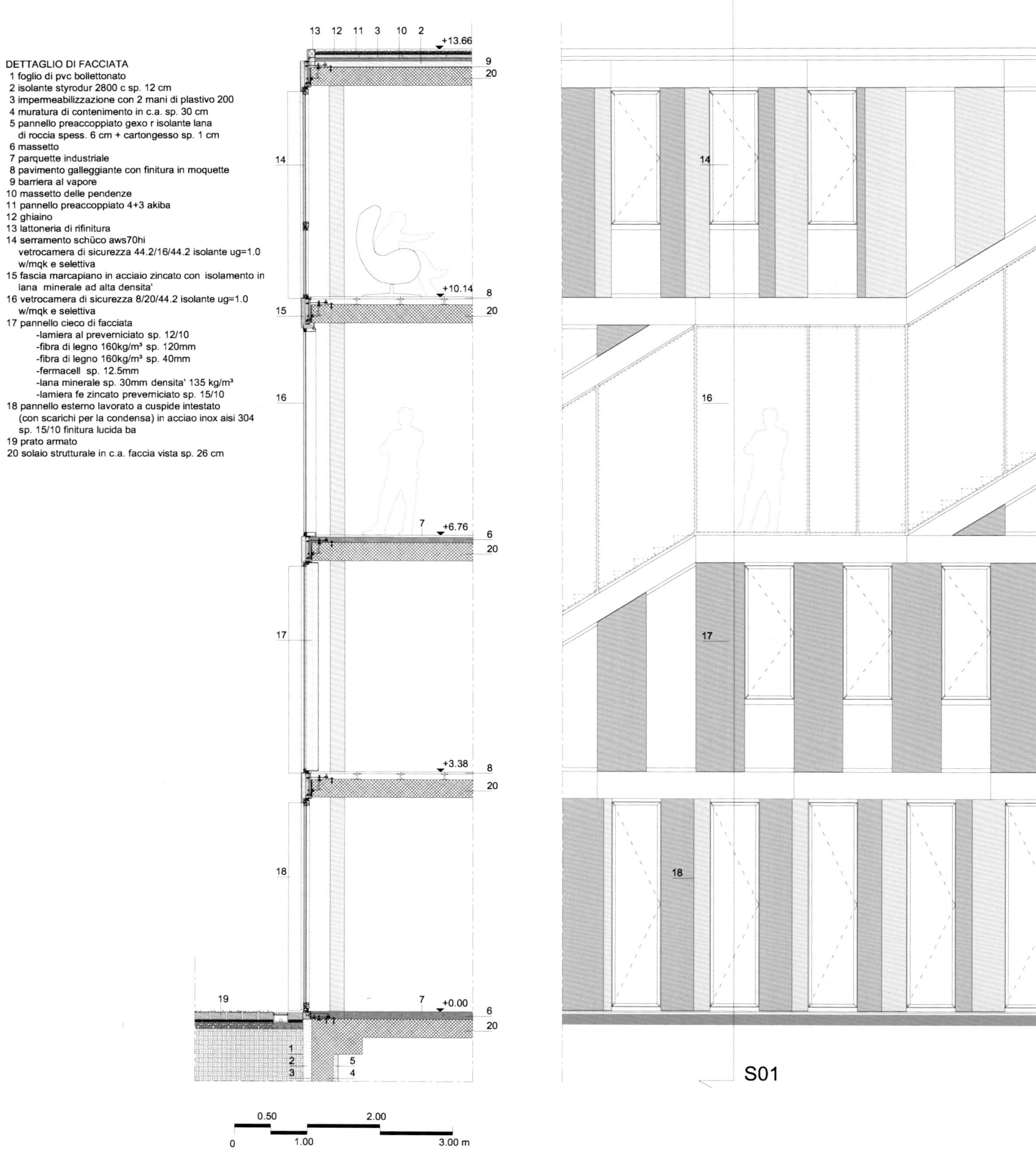

Section Detail

© JULIEN LANOO

© JULIEN LANOO

© SOLANGE SOUZA

© JULIEN LANOO

© SOLANGE SOUZA

© JULIEN LANOO

© JULIEN LANOO

© JULIEN LANOO

DETTAGLIO DI FACCIATA

1 corpo illuminante
2 isolante styrodur 2800 c sp. 12 cm
3 impermeabilizzazione con 2 mani di plastivo 200
4 pannello in lamiera microforata su disegno
5 materassino fonoassorbete in fibra di poliestere sp.30mm densità 40kg/mc
6 lastra di cartongesso da12.5 mm
7 materassino di lana minerale sp. 40/60mm densità 40kg/mq
8 pavimento galleggiante con finitura in moquette
9 barriera al vapore
10 massetto delle pendenze
11 pannello preaccoppiato 4+3 akiba
12 ghiaino
13 lattoneria di rifinitura
14 serramento schüco aws70hi
vetrocamera di sicurezza 44.2/16/44.2 isolante ug=1.0 w/mqk e selettiva
15 fascia marcapiano in acciaio zincato con isolamento in lana minerale ad alta densita'
16 pannello cieco di facciata
-lamiera al preverniciato sp. 12/10
-fibra di legno 160kg/m³ sp. 120mm
-fibra di legno 160kg/m³ sp. 40mm
-fermacell sp. 12.5mm
-lana minerale sp. 30mm densita' 135 kg/m³
-lamiera fe zincato preverniciato sp. 15/10
17 pannello esterno lavorato a cuspide intestato (con scarichi per la condensa) in acciao inox aisi 304 sp. 15/10 finitura lucida ba
18 solaio strutturale in c.a. faccia vista sp. 26 cm

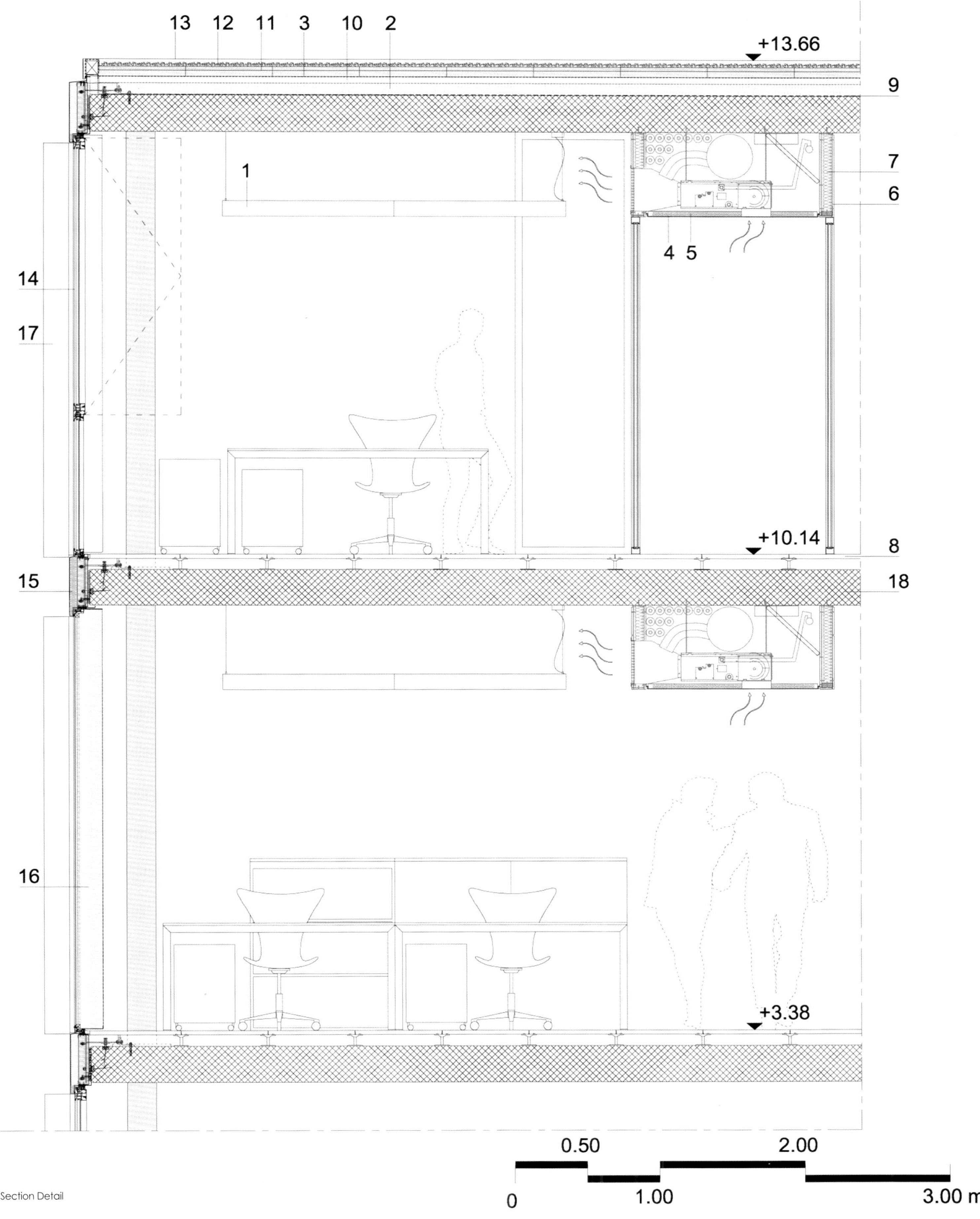

Section Detail

LEROY MERLIN LEROY MERLIN

© SOLANGE SOUZA

interpolis, Tilburg, the Netherlands

NL Architects

© Jeroen Musch

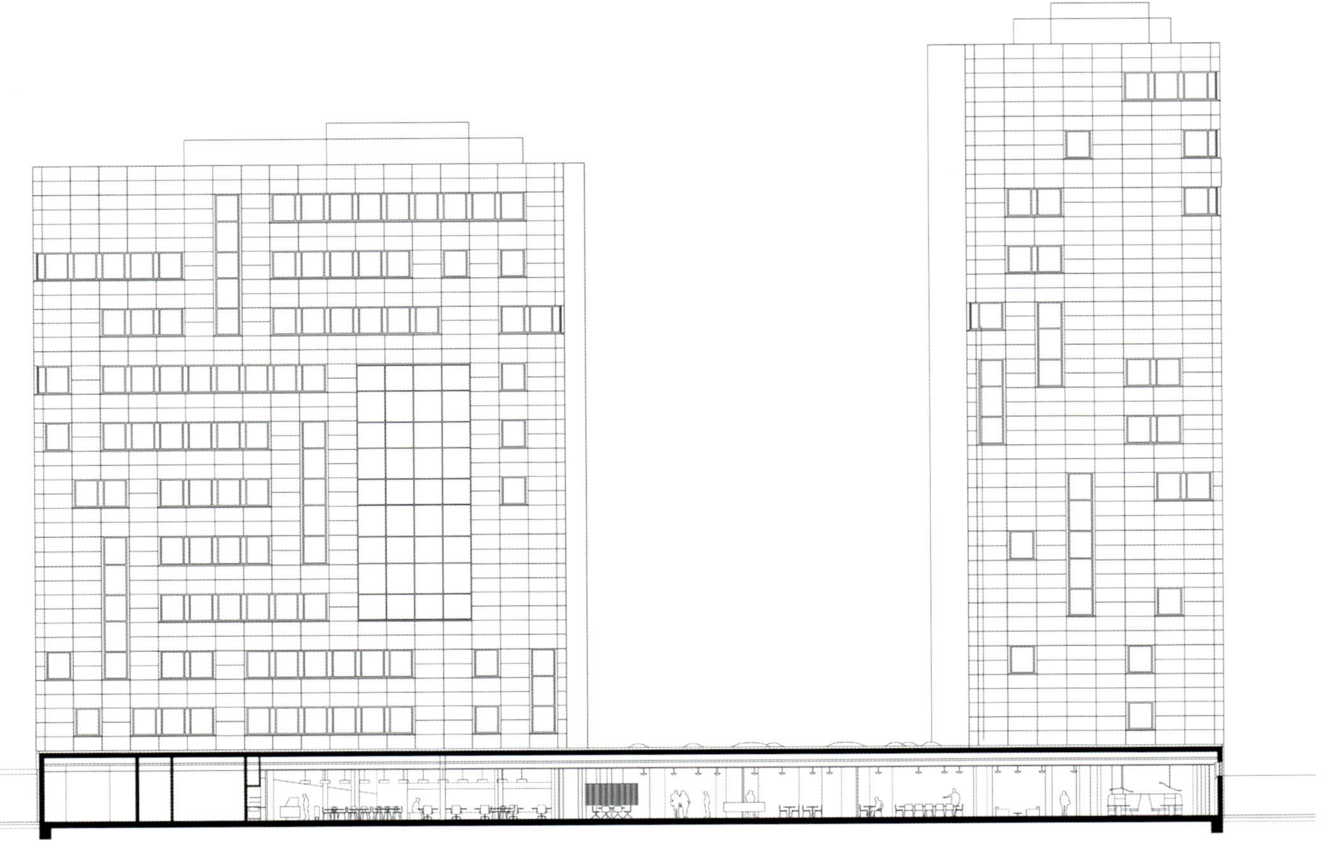

Section

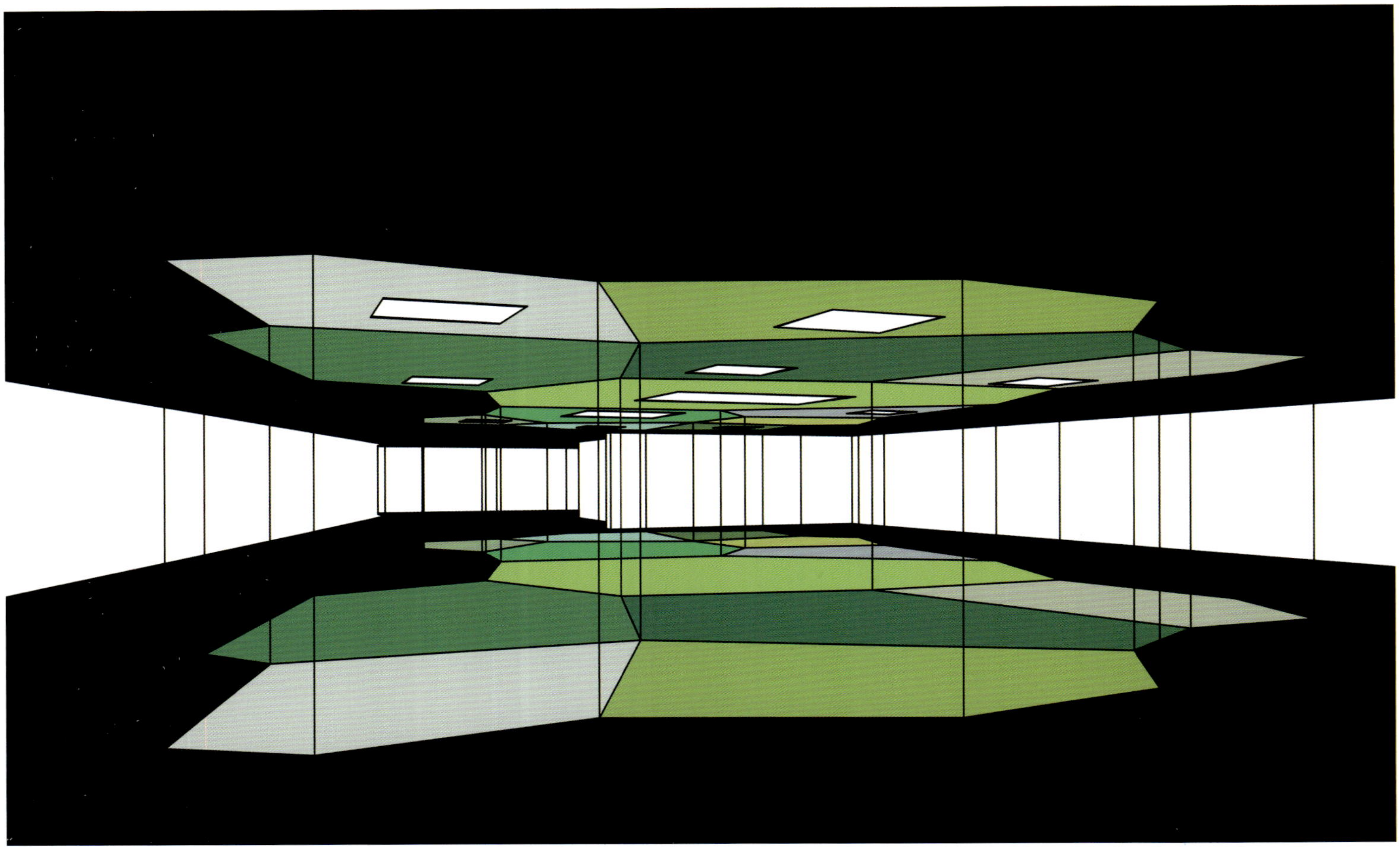

© Jeroen Musch

© Jeroen Musch

Louis Dreyfus Armateurs HEAD QUARTER Suresnes, France

AZC

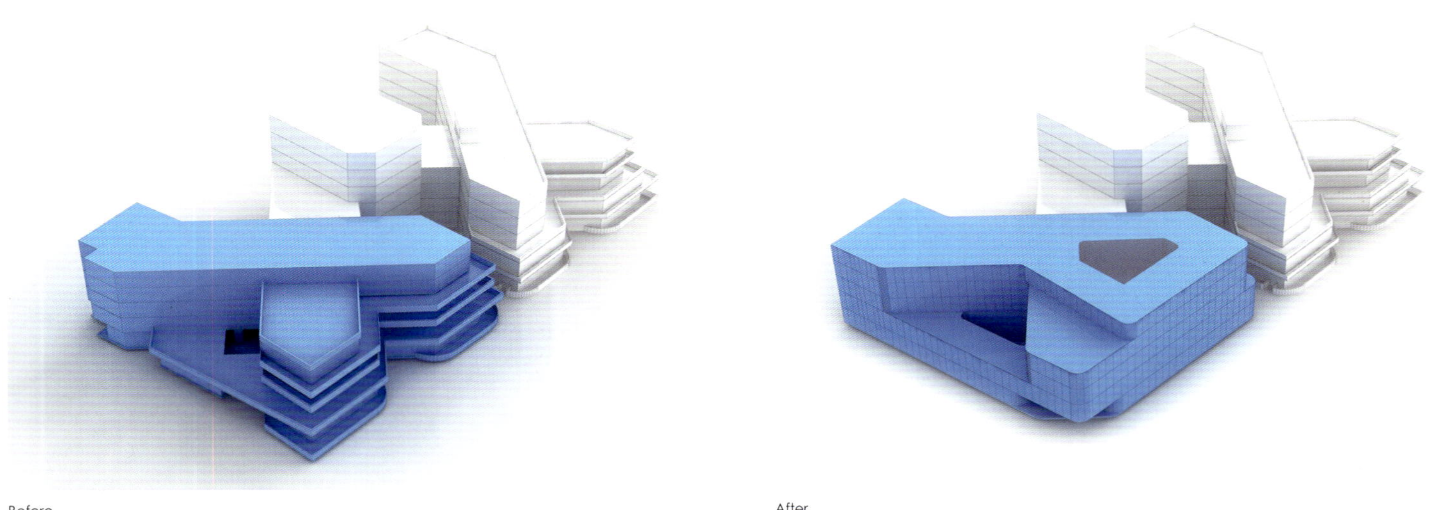

Before

After

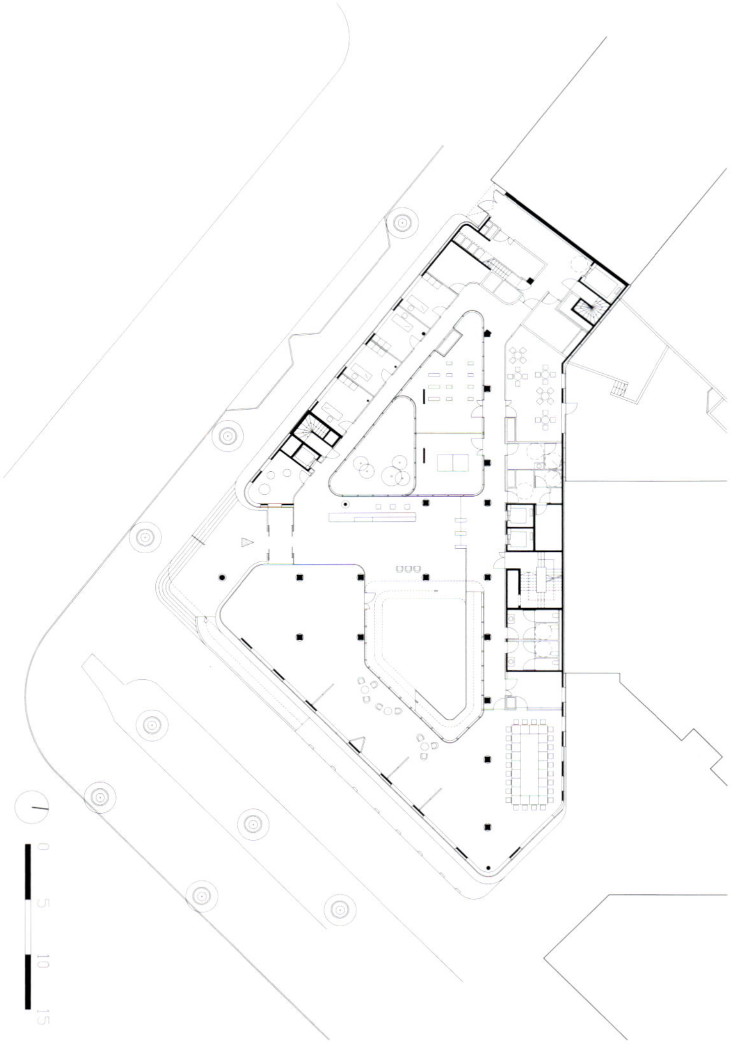

Ground Floor

Fourth Floor

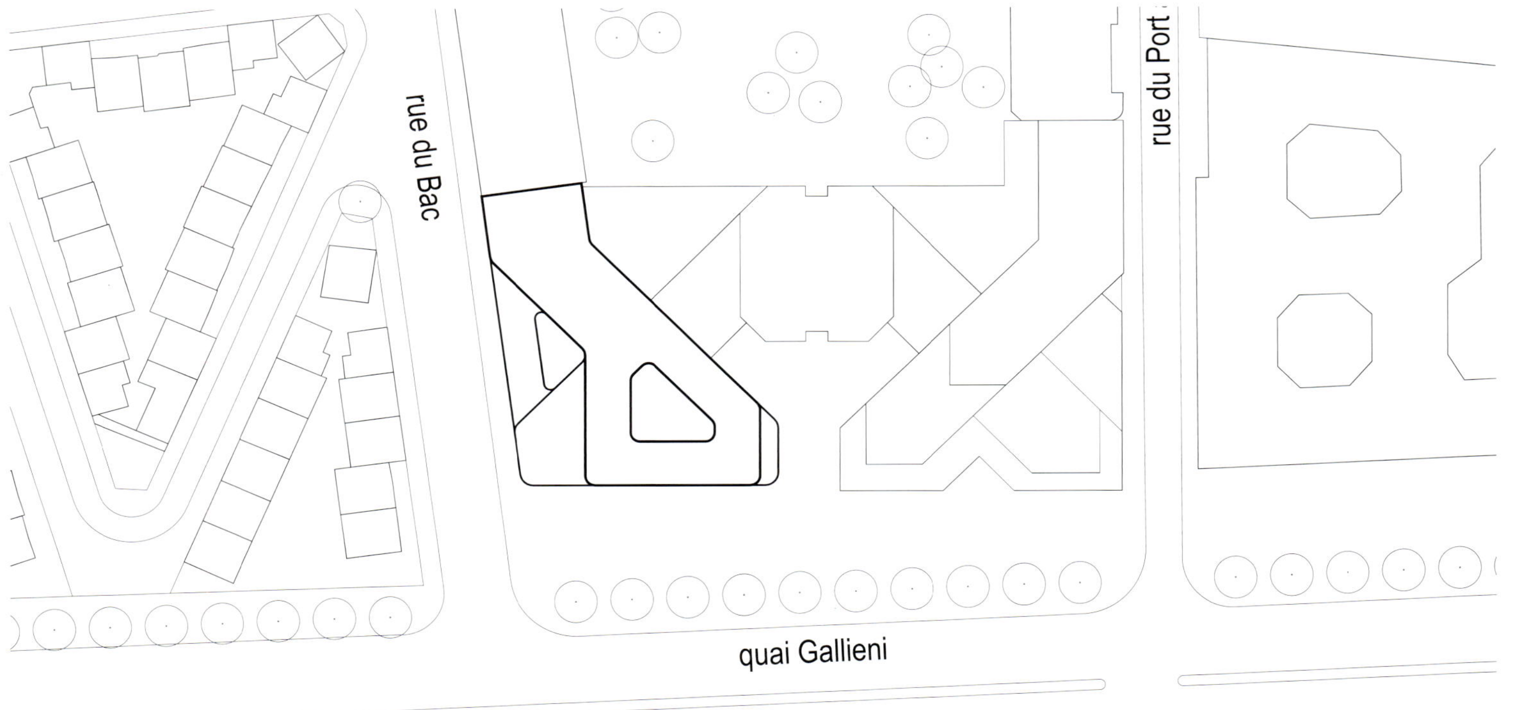

Mass Plan

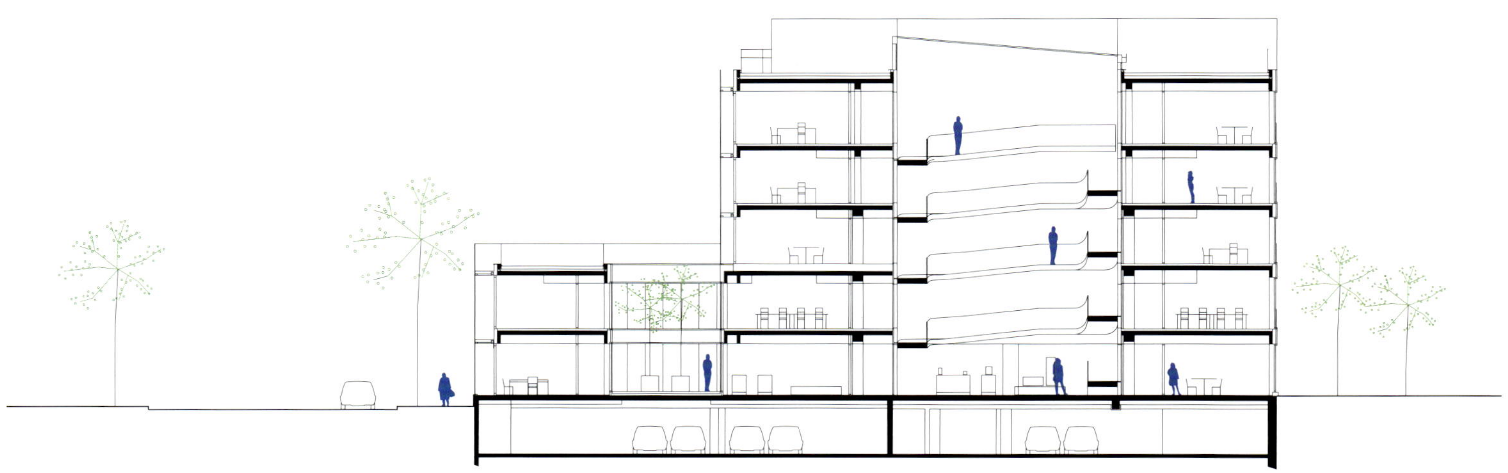

Transverse Section

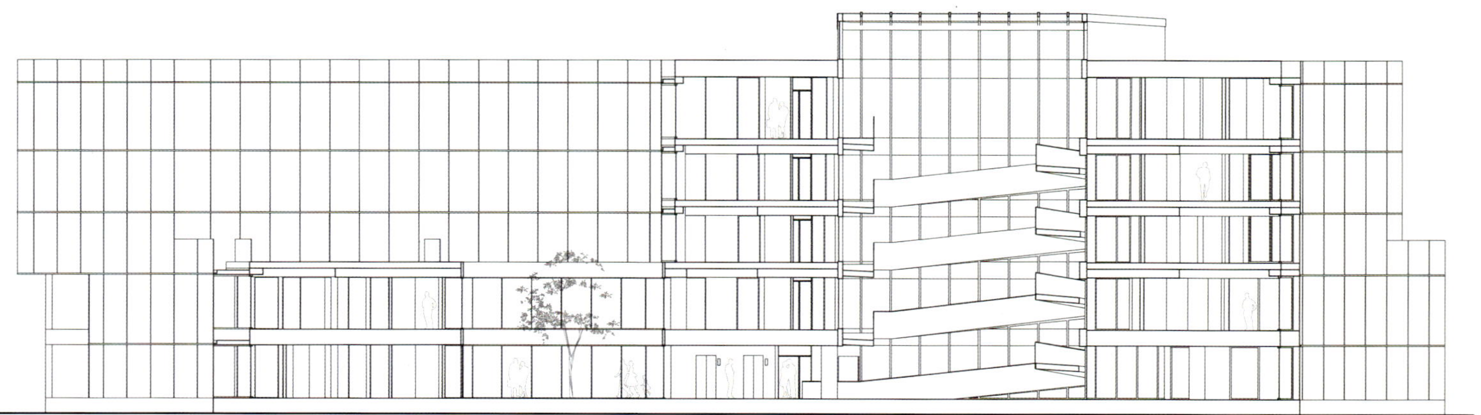

Longitudinal Elevation

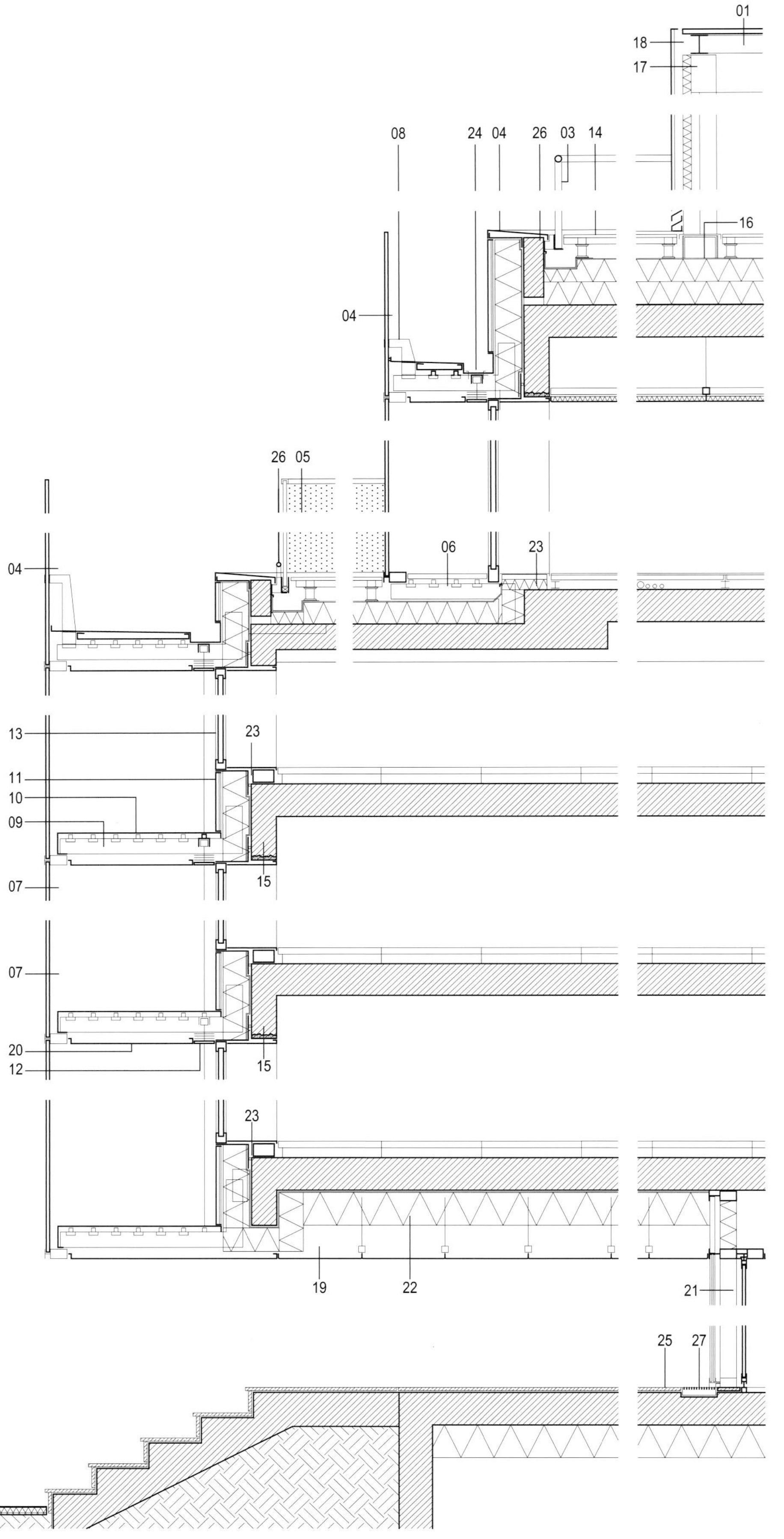

01- CAILLEBOTIS ACIER GALVANISE
02- TOLE ALUMINIUM DE COUVERTURE
02- GARDE CORPS DE SECURITE
04- ECRAN VITRE FIXE, SERIGRAPHIE PARTIELLE
05- GARDE CORPS VERRE
06- DALLES GRANITO
07- PEAU EXTERIEURE :
PROFIL ACIER A CAPOT SERREUR HORIZONTAL
VERRE FEUILLETE EXTRA CLAIR, SERIGRAPHIE
08- STRUCTURE ACIER
09- CONSOLE ACIER
10- PLATEFORME DE MAINTENANCE
11- PEAU INTERIEURE :
HABILLAGE CASSETTE ALUMINIUM EXTRUDE
ISOLATION THERMIQUE LAINE MINERALE
POUTRE / VOILE BETON
12- PROTECTION SOLAIRE, STORE VENETIEN
13- CHASSIS VITRE ALUMINIUM FIXE / OUVRANT
14- TOITURE
PLATELAGE BOIS
DEUX COUCHES ETANCHEITE
ISOLATION THERMIQUE FEUILLUREE COLLEE
PARE VAPEUR
DALLE BETON
15- TOLE ACIER DE FINITION DES EBRASEMENTS
16- PLOT POUR FIXATION TUBE
17- TUBE METALLIQUE D'OSSATURE
18- BARDAGE CASSETTE ALUMINIUM EXTRUDE
19- FAUX PLAFOND SUSPENDU METAL PERFORE
20- TOLE ACIER DE CAPOTAGE
21- CANIVEAU + CAILLEBOTTIS ACIER GALVANISE
22- PORTE VITREE COULISSANTE AUTOMATIQUE
23- ISOLATION THERMIQUE LAINE MINERALE
24- SEUIL TOLE ACIER
25- CHENEAU
26- DALLES GRANITO SUR ETANCHEITE LIQUIDE
27- ACROTERE BETON SUR RUPTEUR THERMIQUE
28- CANIVEAU PLAT
29- CHANFREIN ANGLE RENTRANT

Detail

Airbus on board

21
ACCES
POMPIERS

Port House Antwerp, Antwerp, Belgium

Zaha Hadid Architects

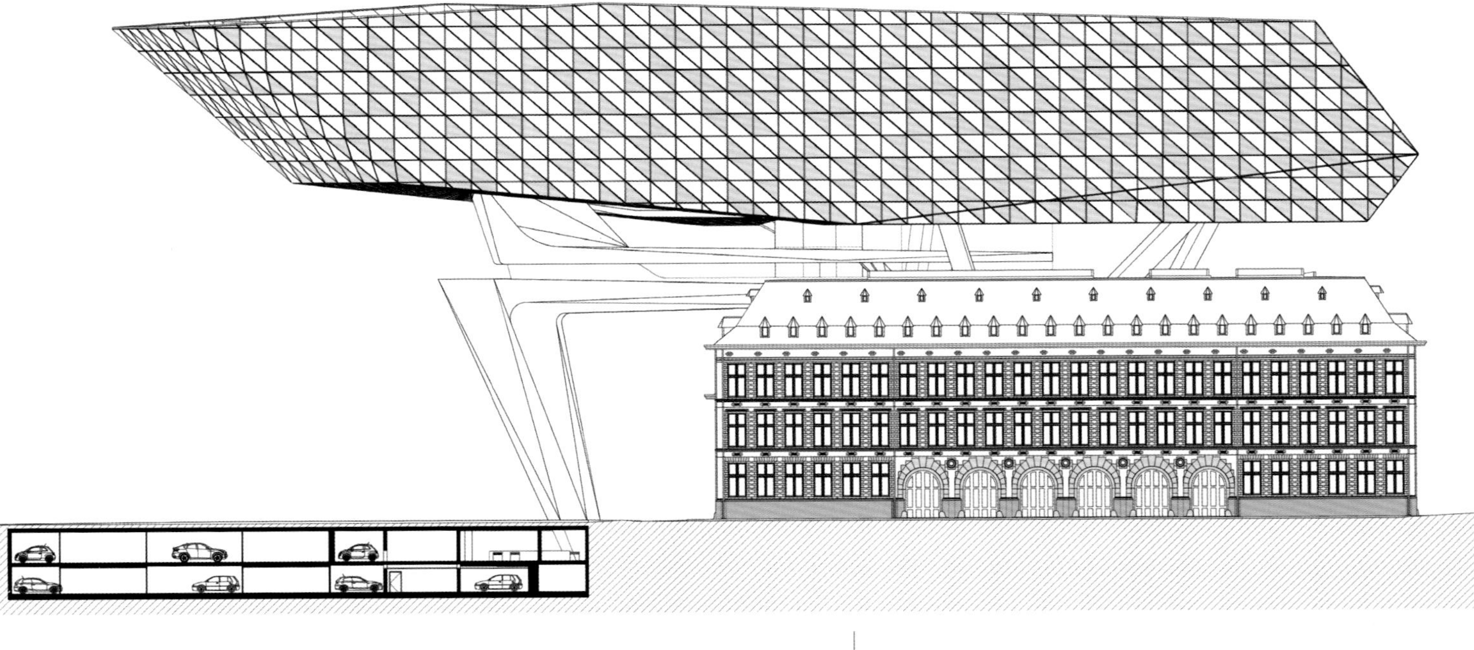

East Elevation

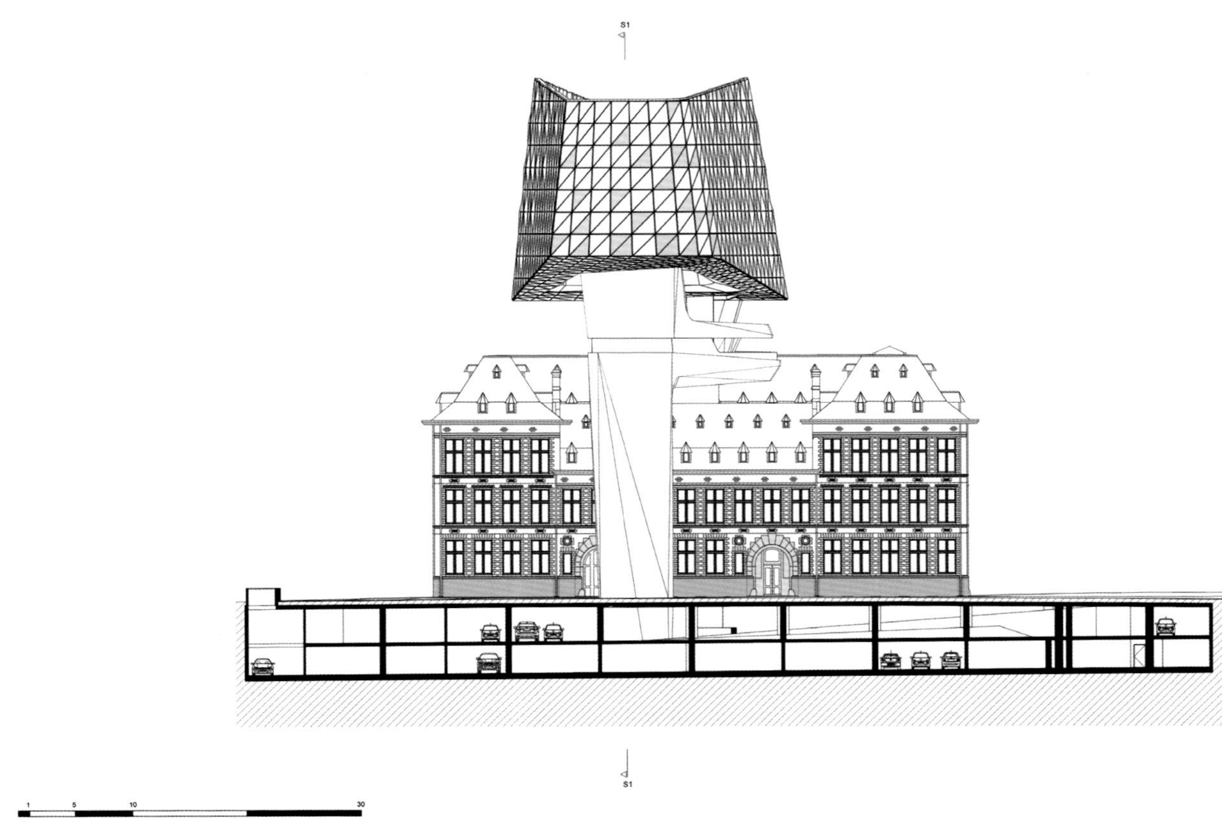

South Elevation

© Tim Fisher

© Hufton+Crow

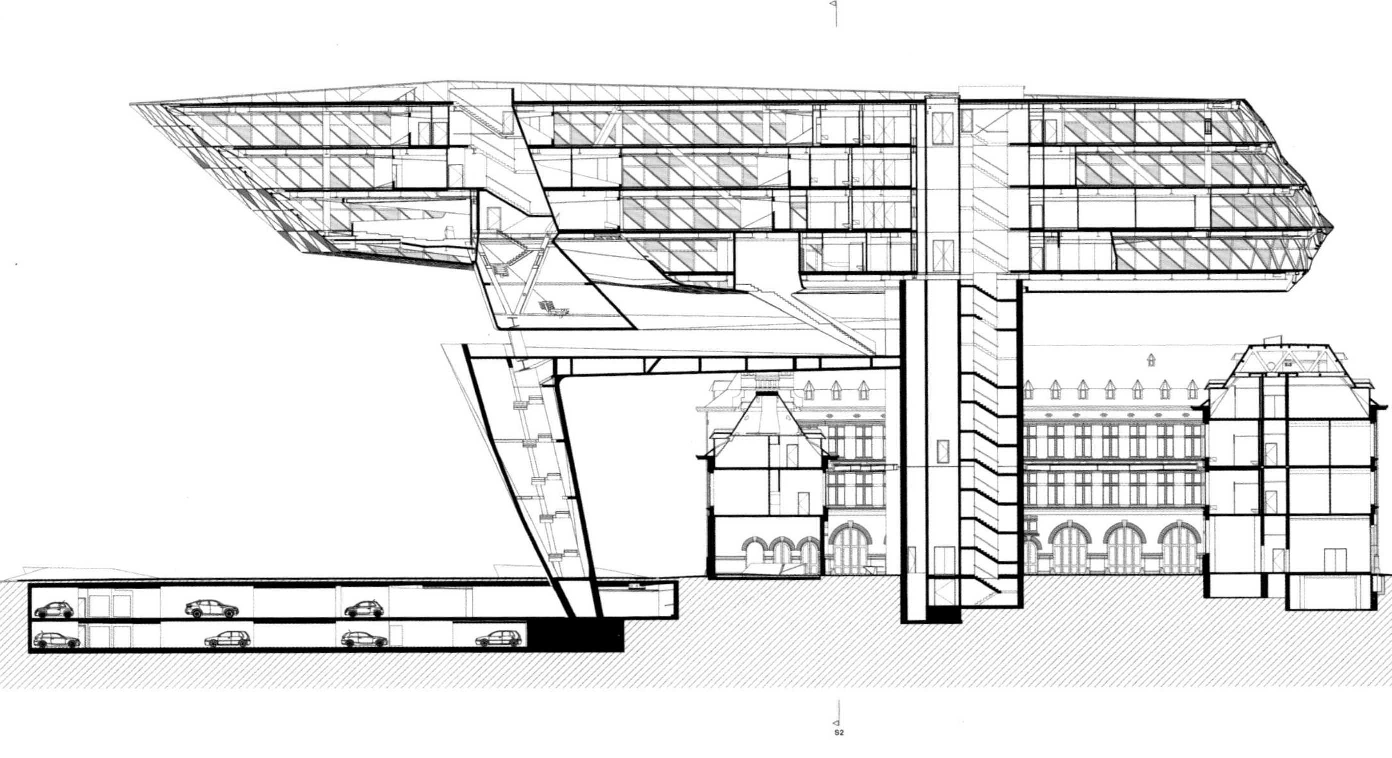

Longitudinal Section

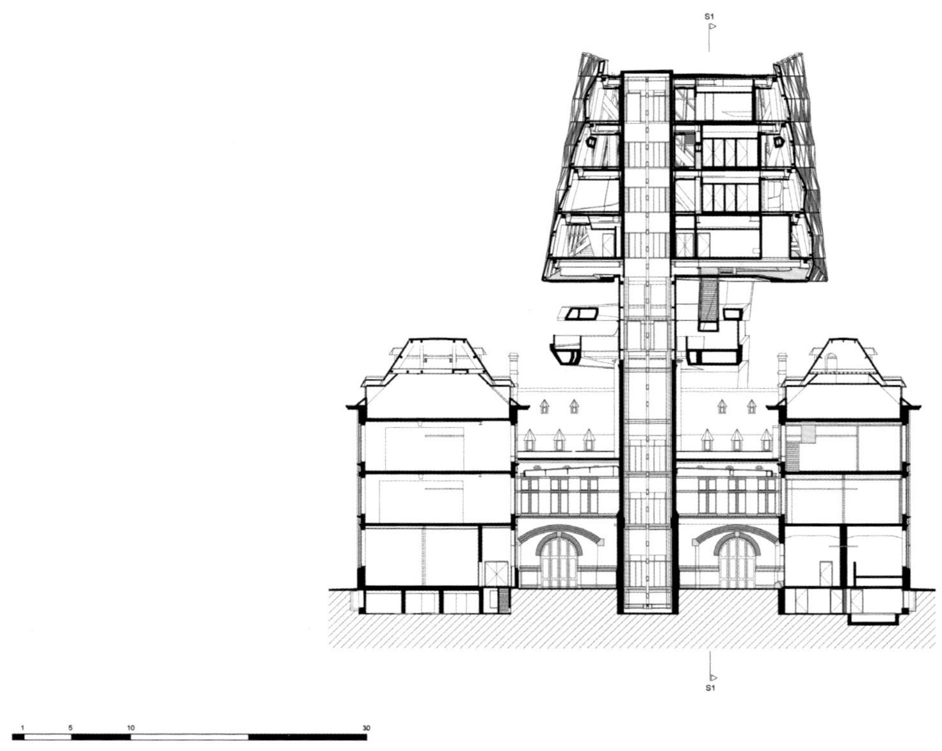

Cross Section

© Tim Fisher

© Tim Fisher

© Tim Fisher

© Tim Fisher

© Hufton+Crow

© Tim Fisher

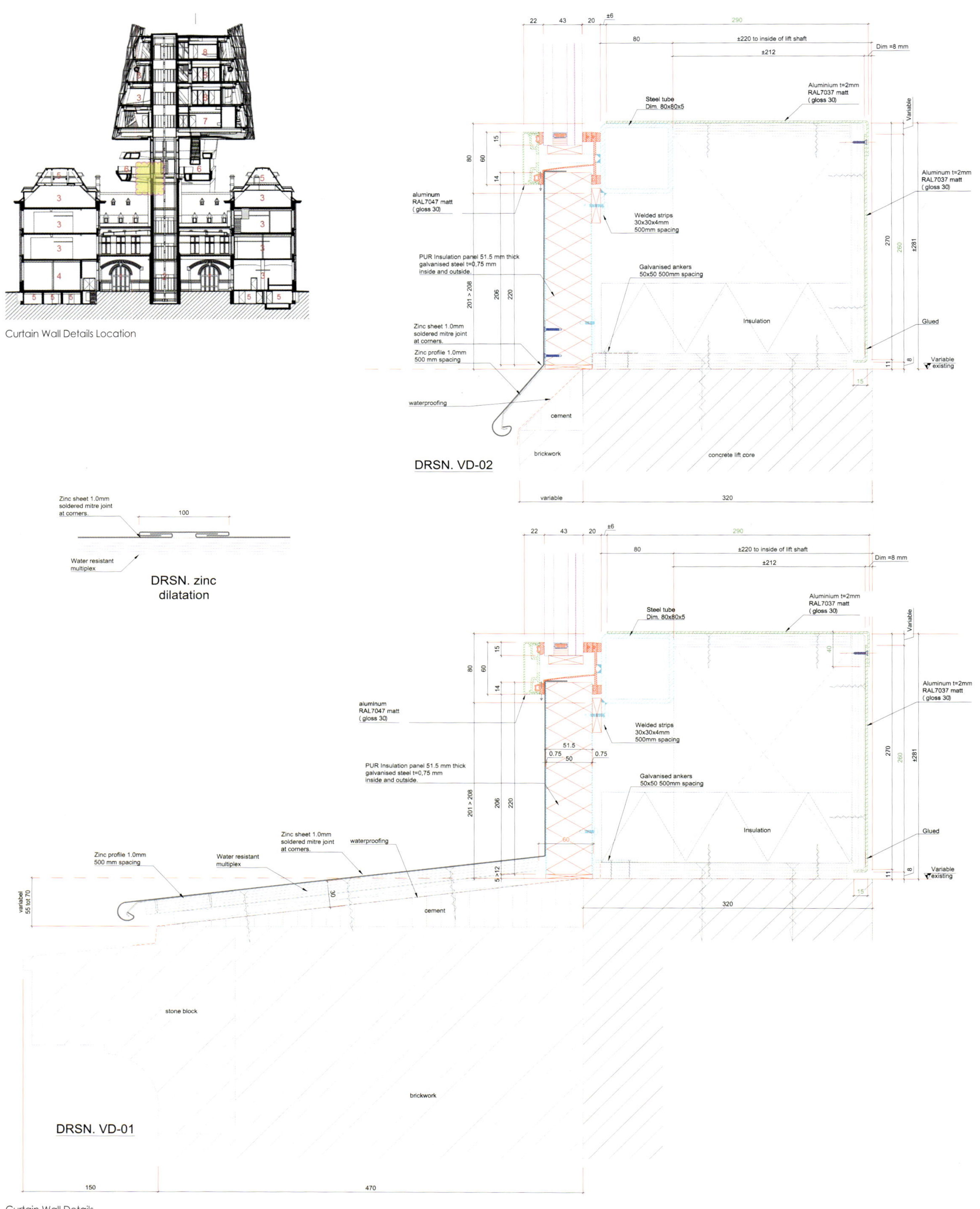

Curtain Wall Details

© Tim Fisher

© HeleneBinet

© Hufton+Crow

RATP BUS CENTRE, Thiais, France
ECDM

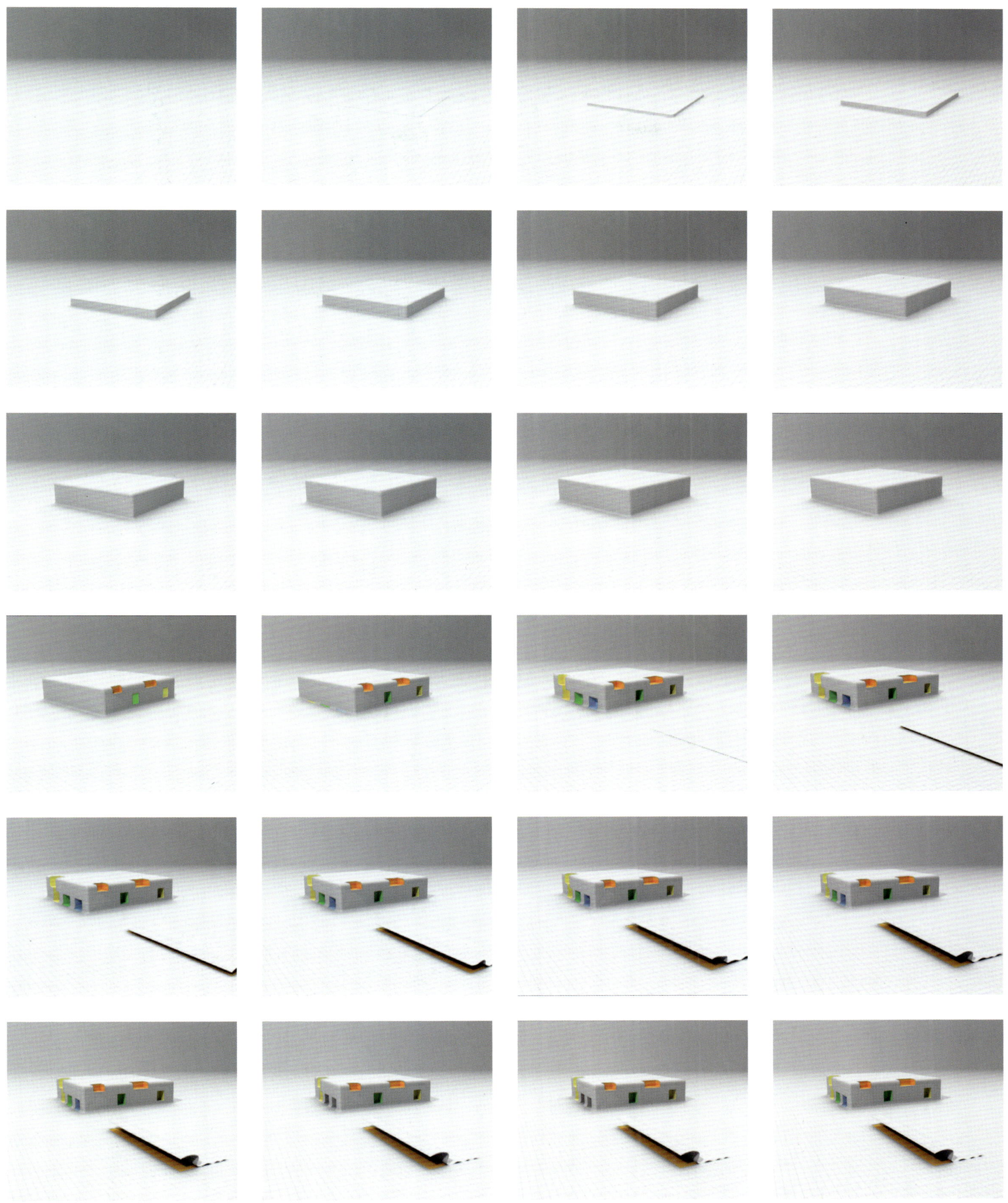

Study Modeling

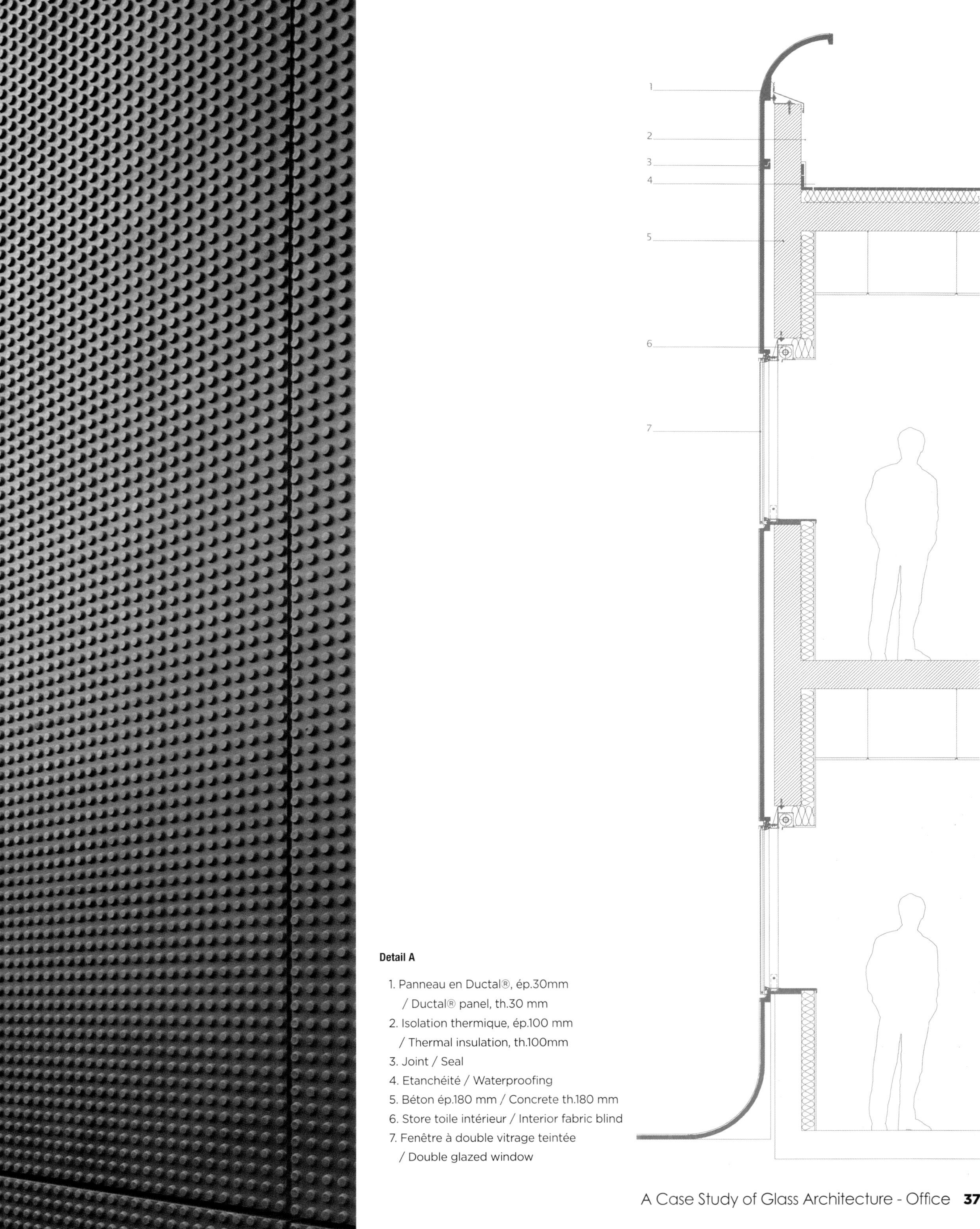

Detail A

1. Panneau en Ductal®, ép.30mm / Ductal® panel, th.30 mm
2. Isolation thermique, ép.100 mm / Thermal insulation, th.100mm
3. Joint / Seal
4. Etanchéité / Waterproofing
5. Béton ép.180 mm / Concrete th.180 mm
6. Store toile intérieur / Interior fabric blind
7. Fenêtre à double vitrage teintée / Double glazed window

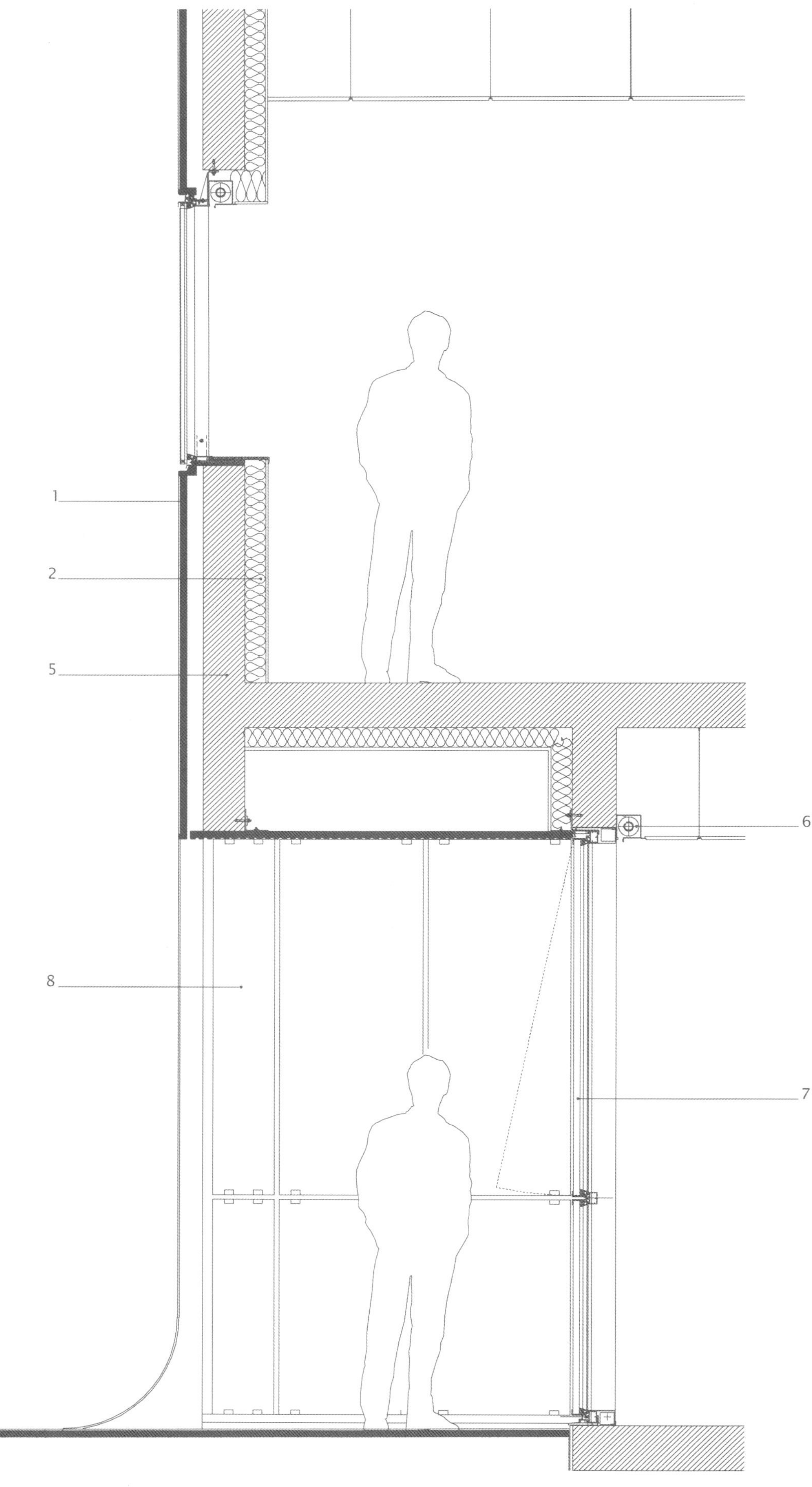

Detail B

1. Panneau en Ductal® / Ductal® panel
2. Isolation thermique, ép.100 mm / Thermal insulation, th.100 mm
3. Joint / Seal
4. Etanchéité / Waterproofing
5. Béton, ép.180 mm / Concrete, th.180 mm
6. Store toile intérieur / Interior fabric blind
7. Fenêtre à double vitrage teintée / Double colourde glazed window
8. Vitrage sur beton coloré / Glazing on coloured concrete

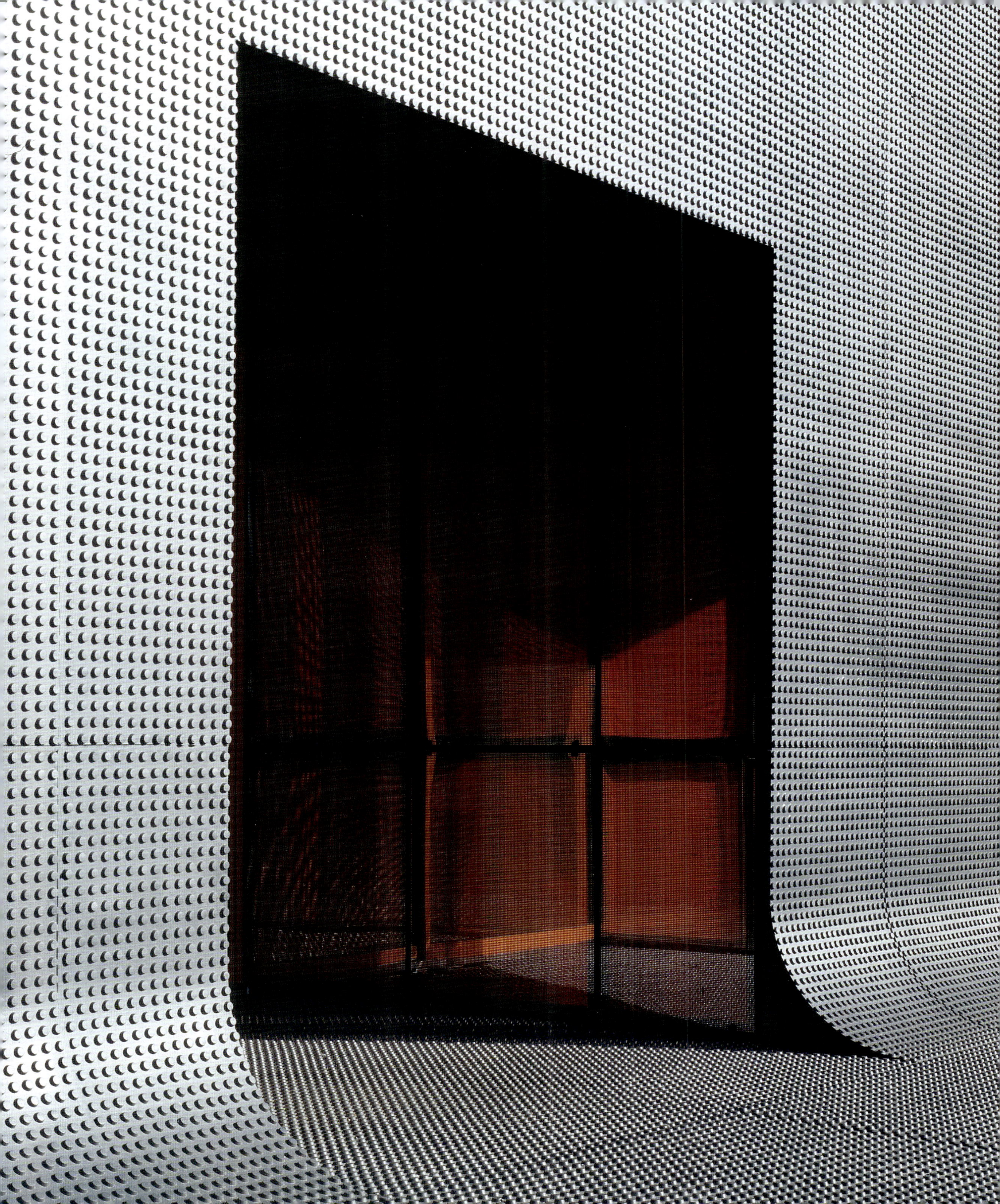

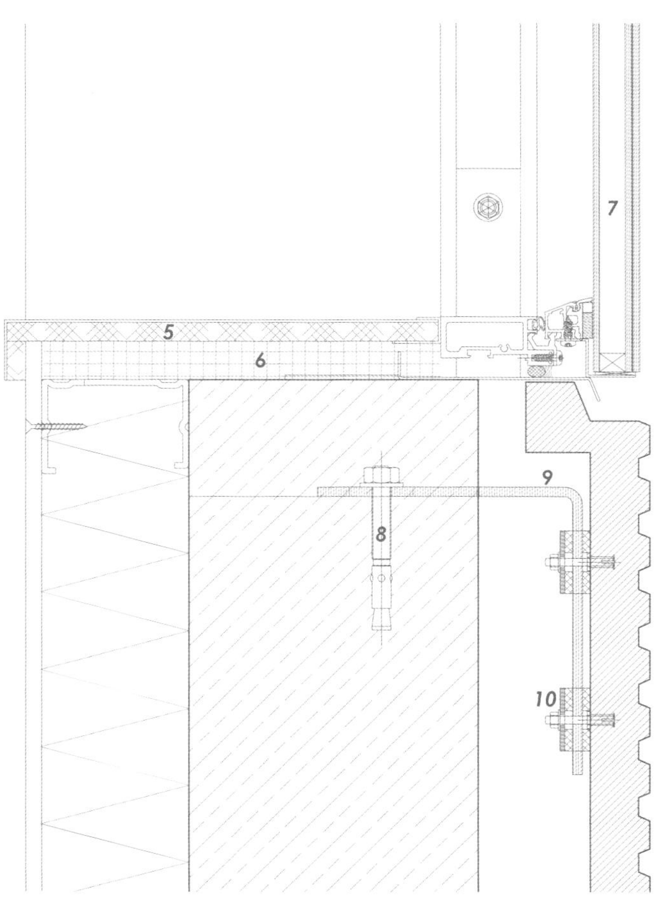

Detail of junction of Ductal® and glazing scale 1:5

1. Panneau Ductal®, ép.30 mm / Ductal® panel, th.30 mm
2. Isolation thermique, ép.90 mm / Thermal insulation, th.90 mm
3. Etanchéité / Waterproofing
4. Plaque de plâtre / Plasterboard
5. Tablette bois mélaminé gris argent / Wooden shelf in silver gray melamine
6. Béton, pente 1.5% / Concrete, gradient 1.5%
7. Fenêtre à double vitrage / Double glazed window
8. Cheville Inox, Ø 12x100 mm / Stainless steel screw anchor, Ø 12x100 mm
9. Equerre Inox, ép.6 mm / Stainless steel bracket, th.6 mm
10. Tige soudée sur équerre P1, 16x130 mm / Rod welded on P1 bracket, 16x130 mm
11. Store toile intérieur / Interior fabric blind

Tower for Regione Piemonte, Turin, Italy
FUKSAS

Idea

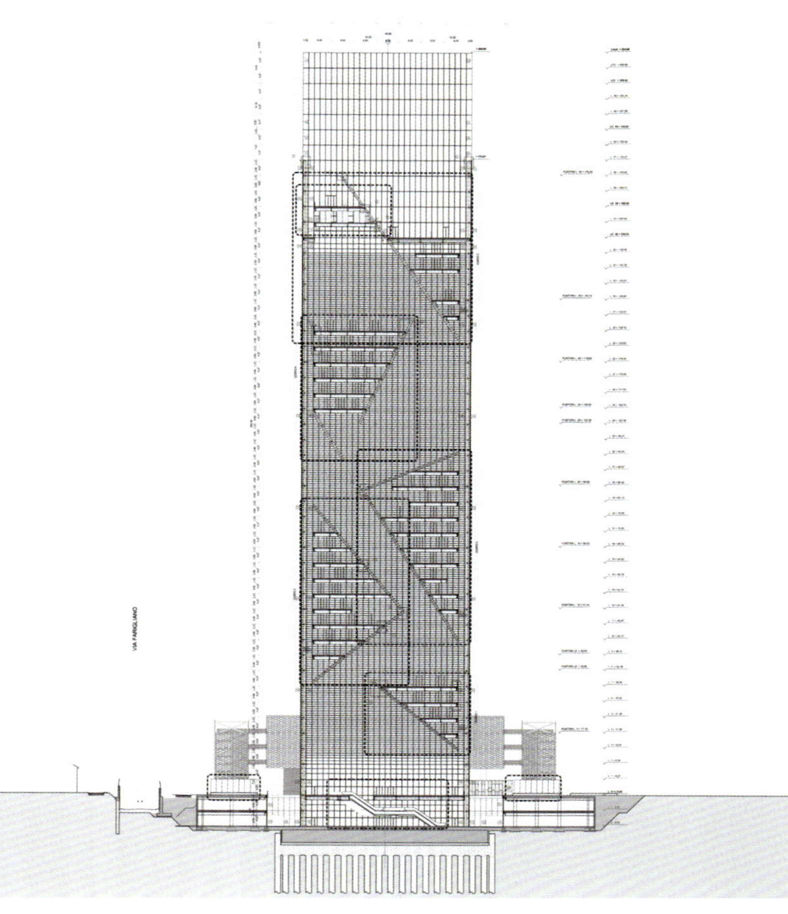
Facade

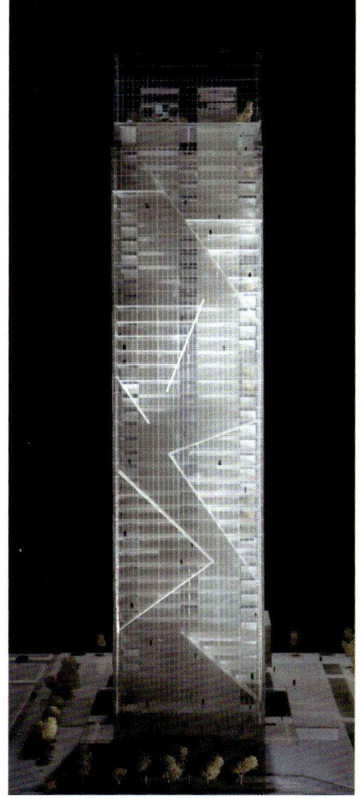

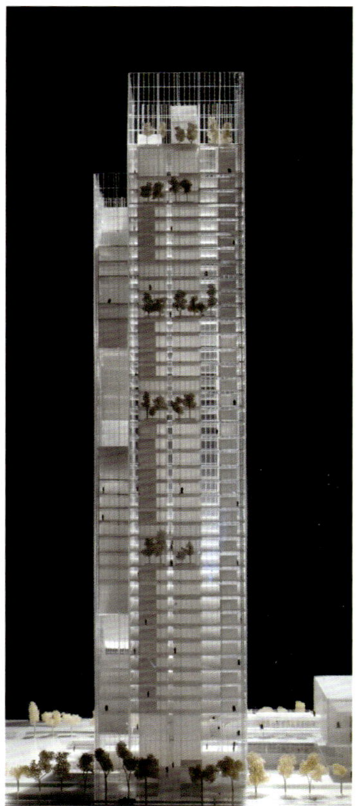

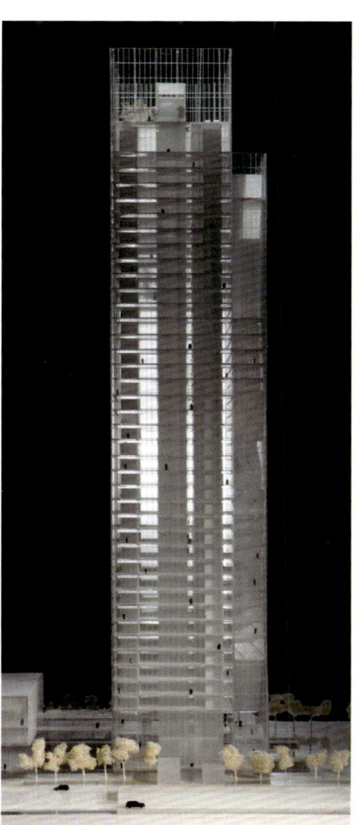

Study Modeling

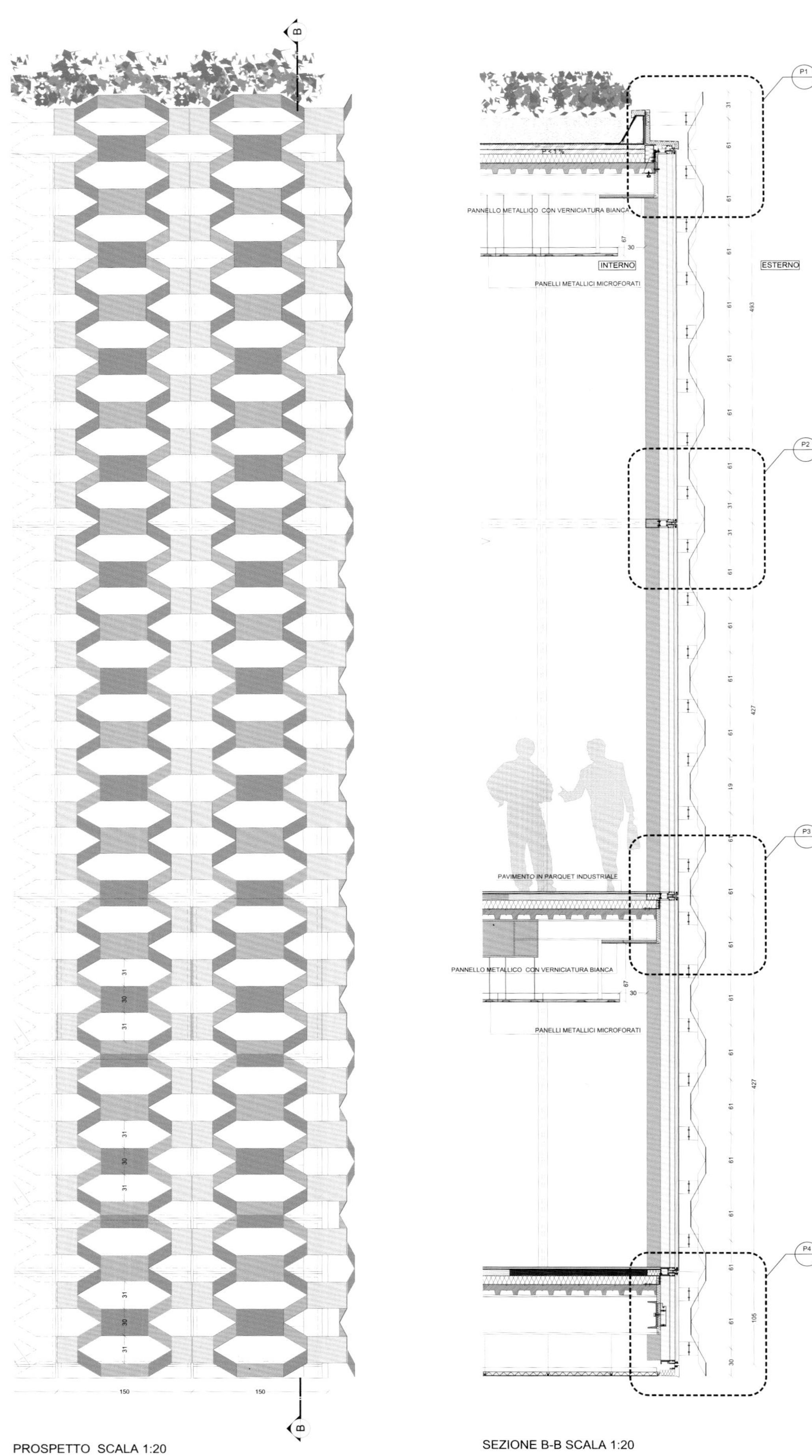

PROSPETTO SCALA 1:20

SEZIONE B-B SCALA 1:20

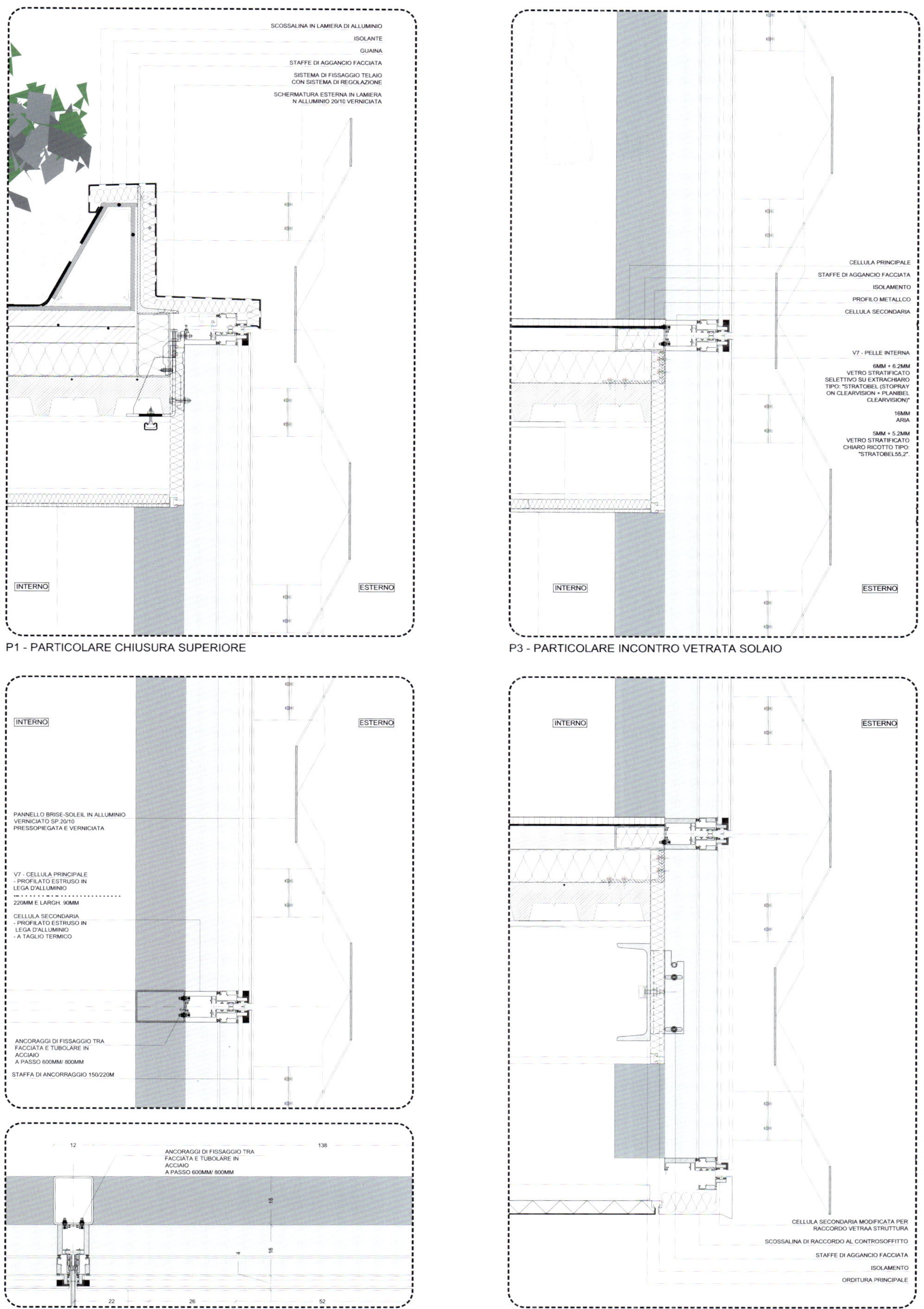

P1 - PARTICOLARE CHIUSURA SUPERIORE

P3 - PARTICOLARE INCONTRO VETRATA SOLAIO

P2 - PARTICOLARE VETRATA A DOPPIA ALTEZZA

P1 - PARTICOLARE CHIUSURA INFERIORE

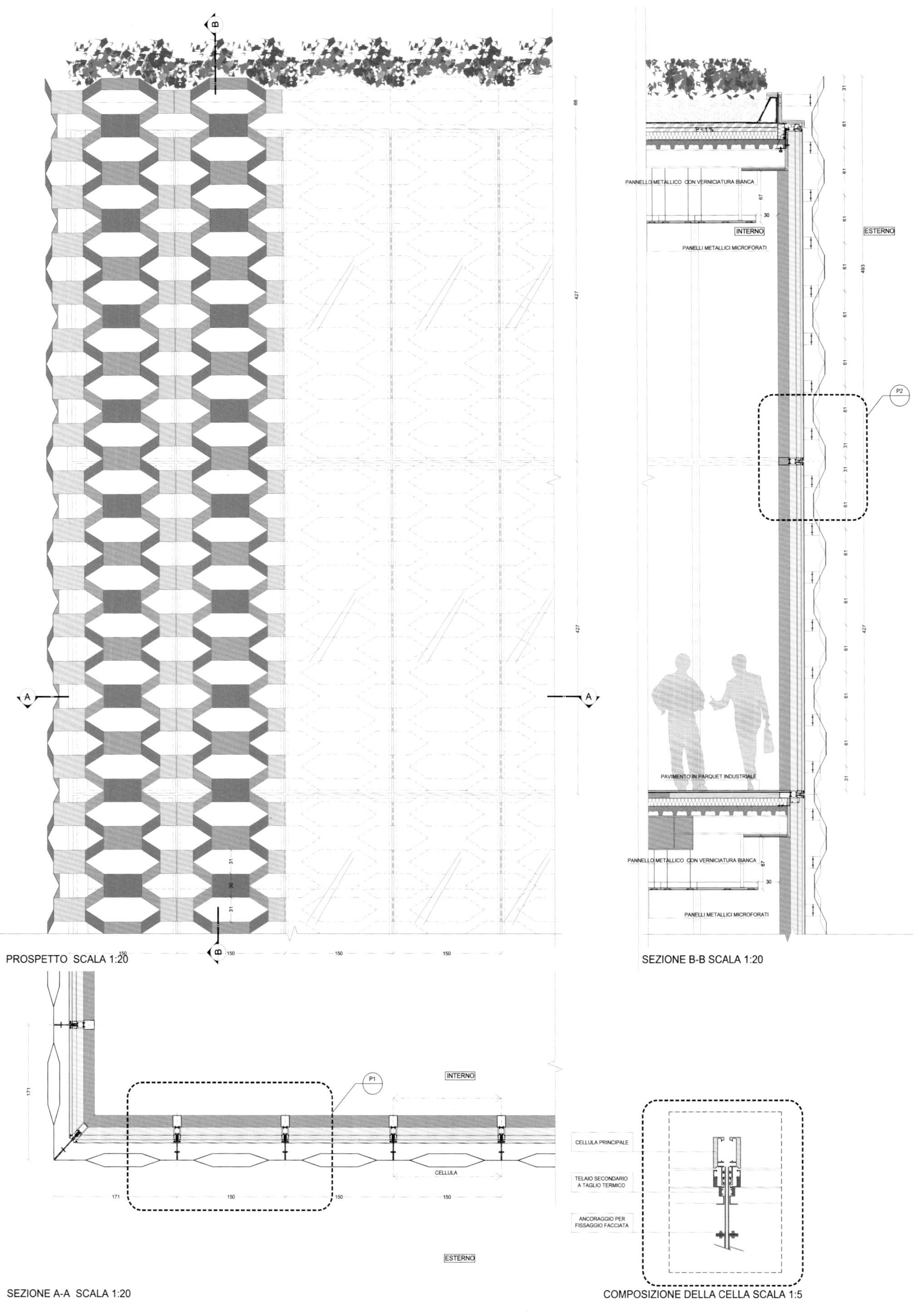
PANNELLO METALLICO CON VERNICIATURA BIANCA
INTERNO
ESTERNO
PANELLI METALLICI MICROFORATI
P2
PAVIMENTO IN PARQUET INDUSTRIALE
PANNELLO METALLICO CON VERNICIATURA BIANCA
PANELLI METALLICI MICROFORATI
PROSPETTO SCALA 1:20
SEZIONE B-B SCALA 1:20
P1
INTERNO
CELLULA
ESTERNO
CELLULA PRINCIPALE
TELAIO SECONDARIO A TAGLIO TERMICO
ANCORAGGIO PER FISSAGGIO FACCIATA
SEZIONE A-A SCALA 1:20
COMPOSIZIONE DELLA CELLA SCALA 1:5

© Andrea Revello

© Archivio Fuksas

© Archivio Fuksas

Cityscope,

marco hemmerling architecture design

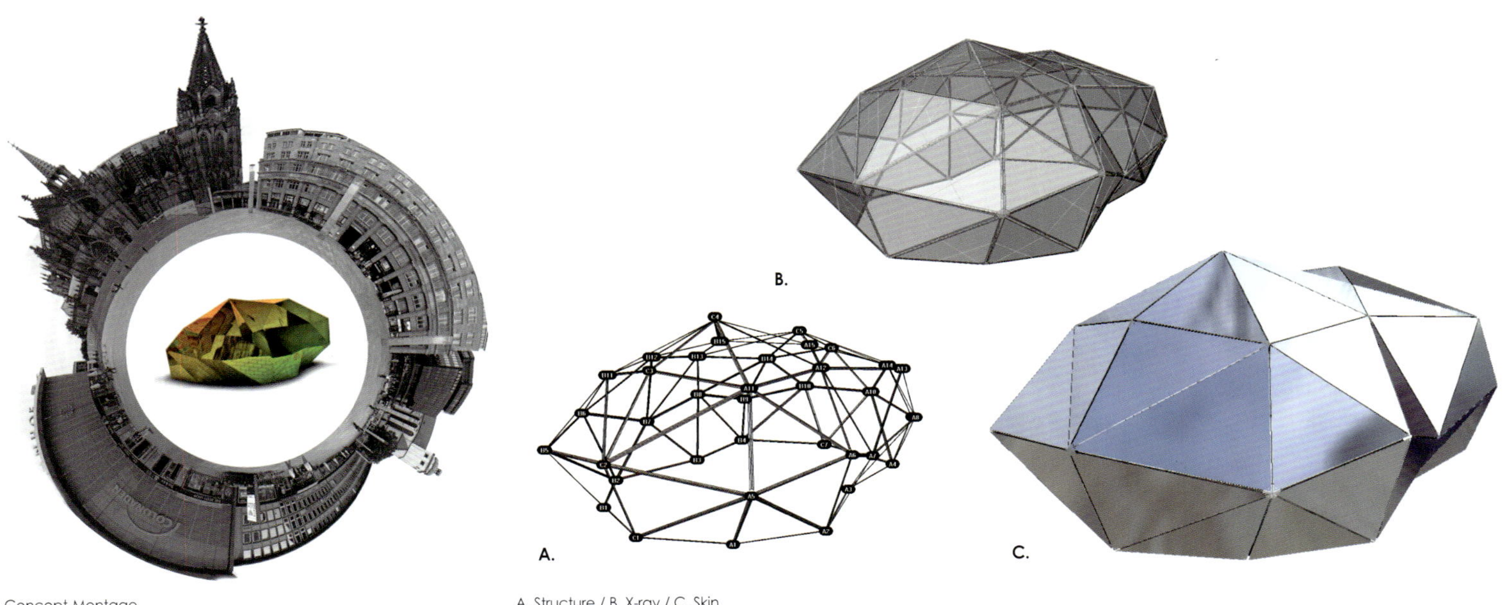

Concept-Montage

A. Structure / B. X-ray / C. Skin

HOF
DB
ZEITCAFE

On the Rocks, Vienna, Austria
the next ENTERprise

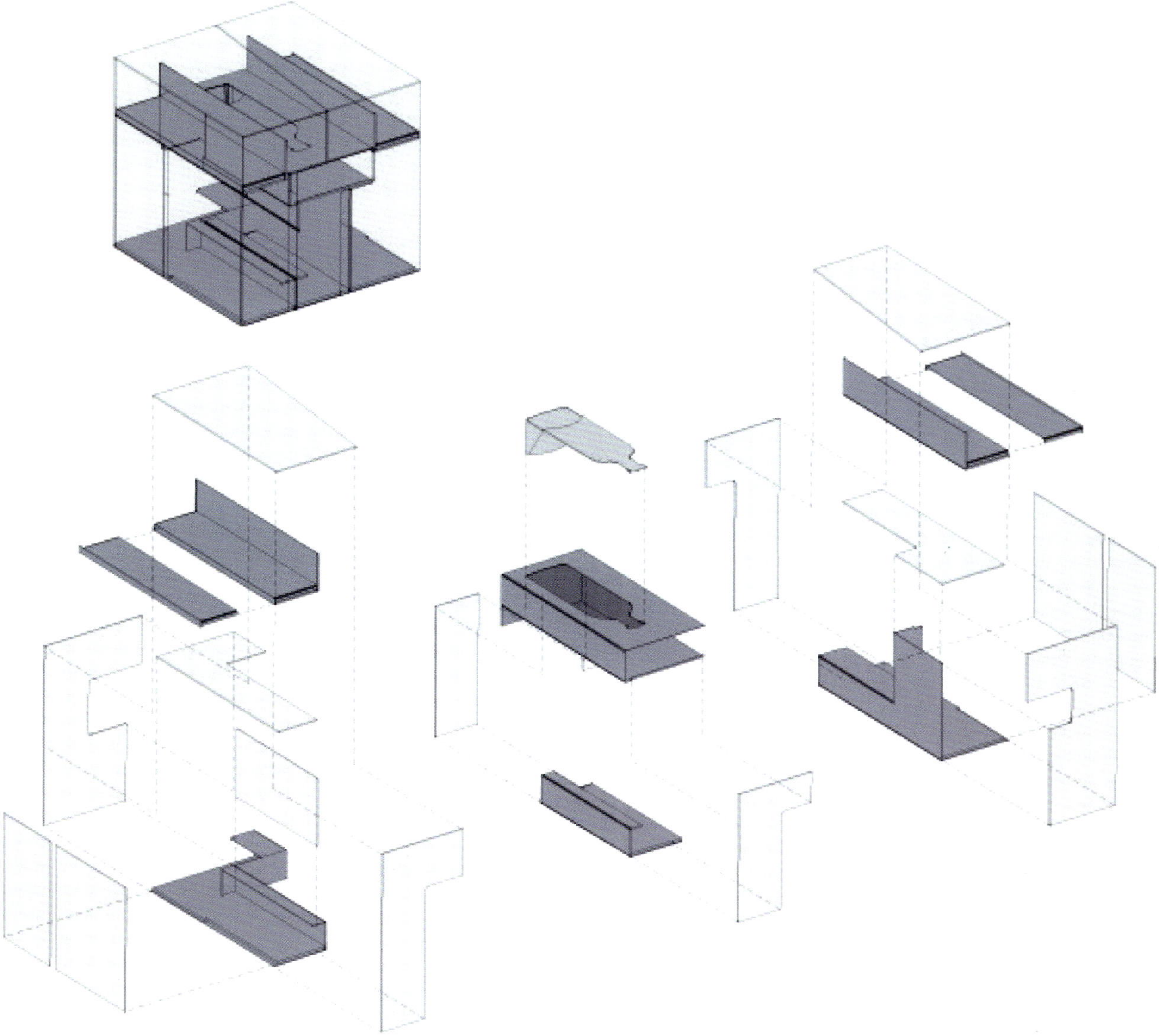

Glass and aluminum elements

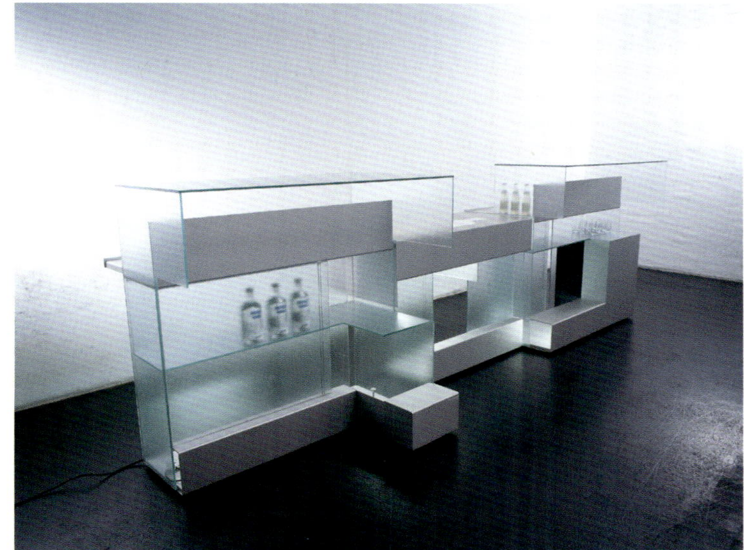

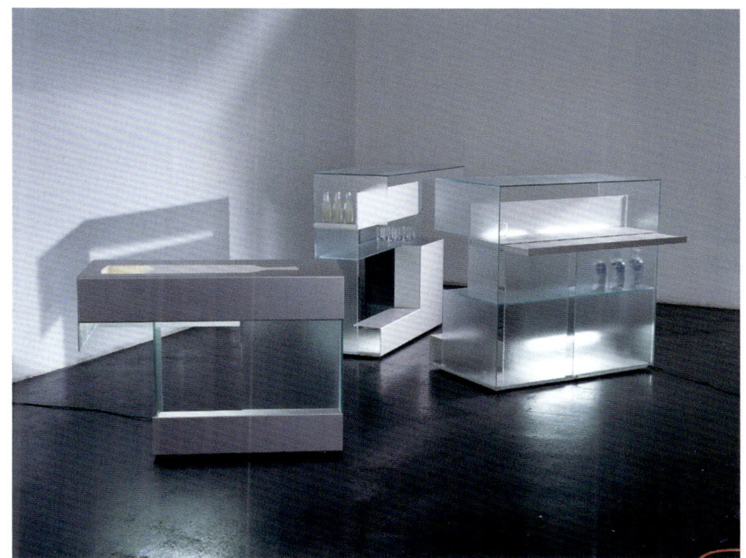

Images showing the different configurations.

Duncan Terrace, London, UK

DOSarchitects Ltd

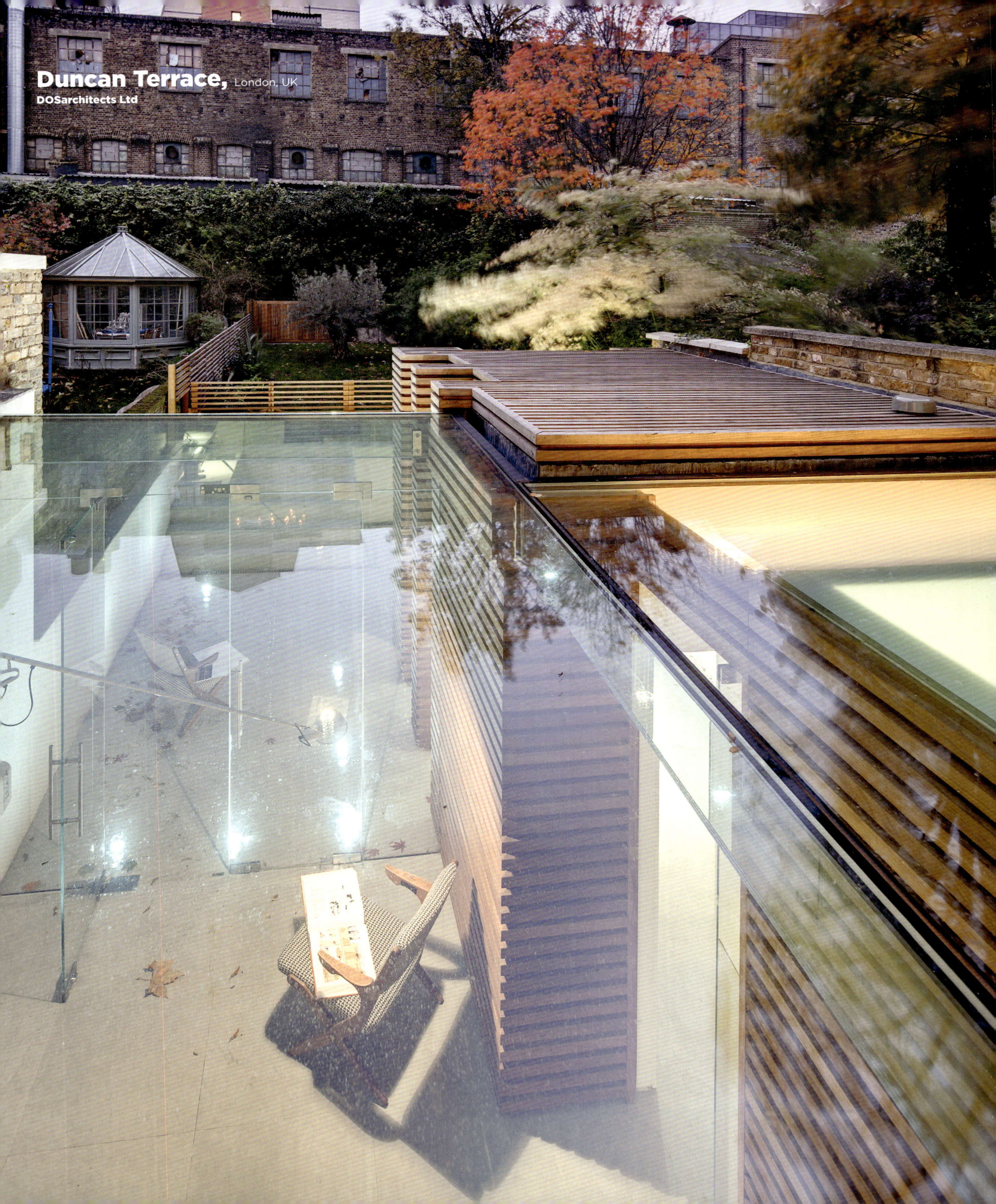

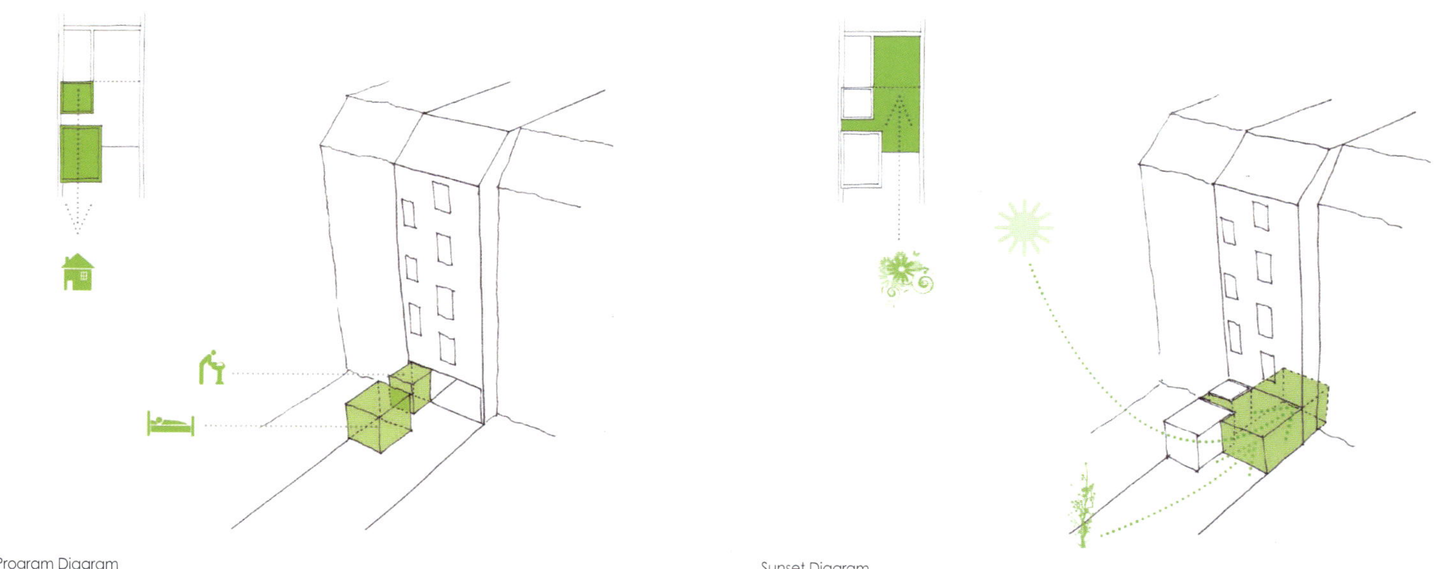

Program Diagram

Sunset Diagram

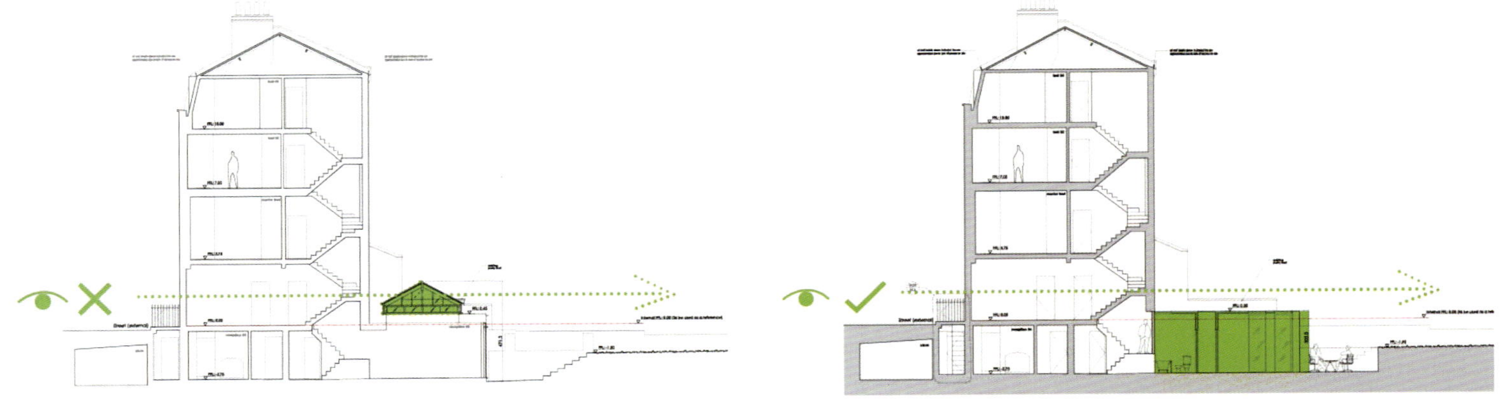
Before → After

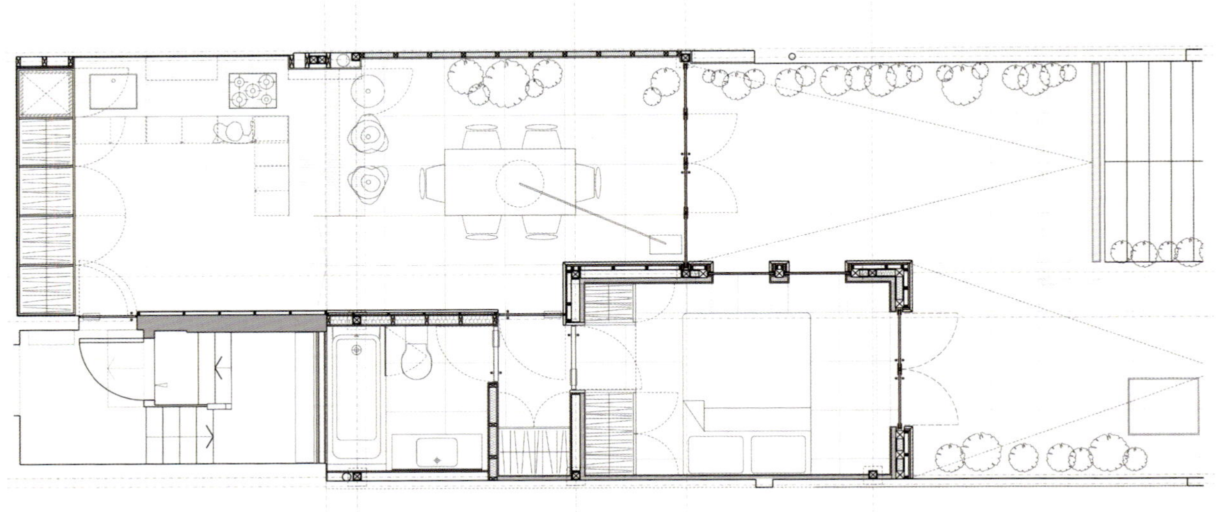
Floor Plan

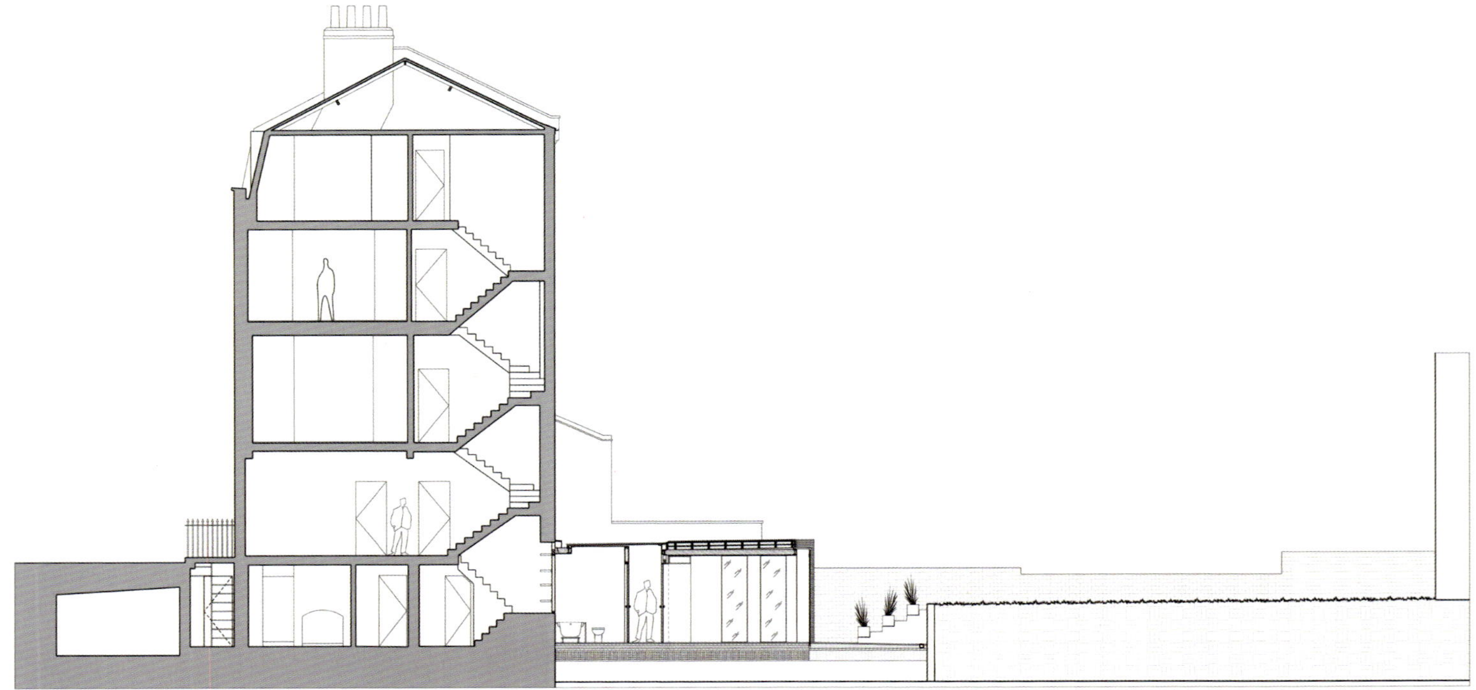
Section

Rivas Undergroun Station, Madrid, Spain

LANDÍNEZ+REY arquitectos

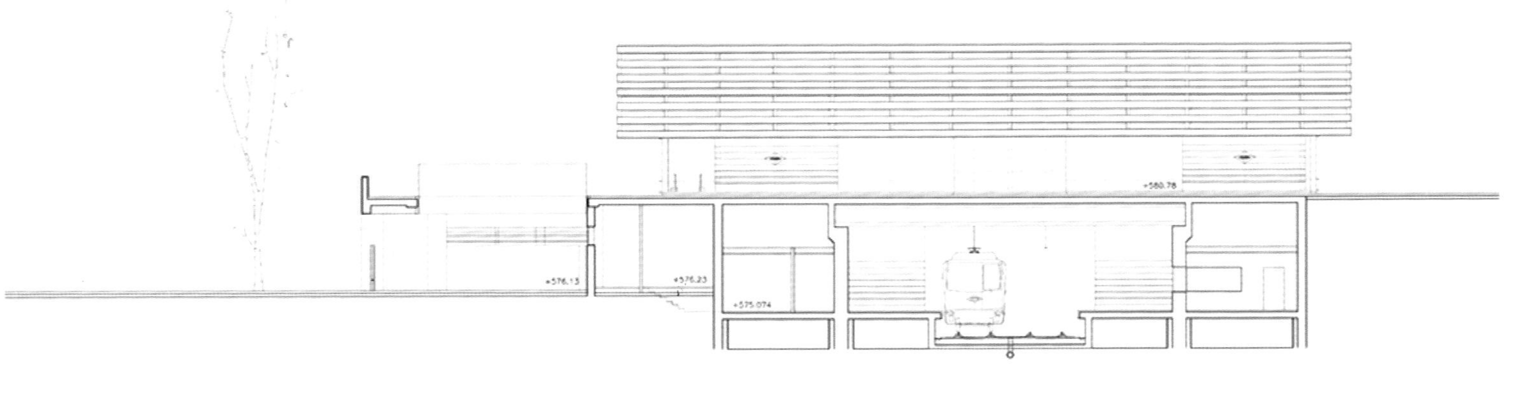

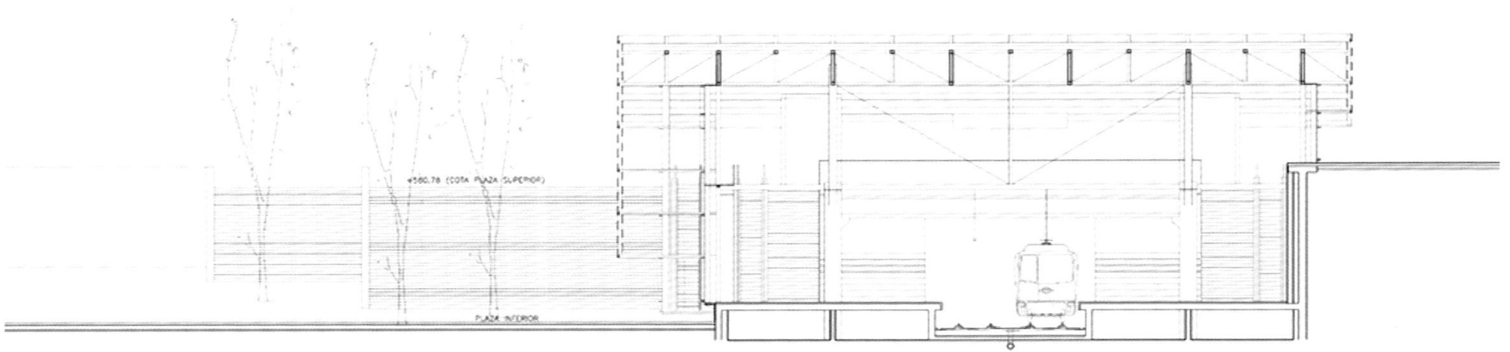

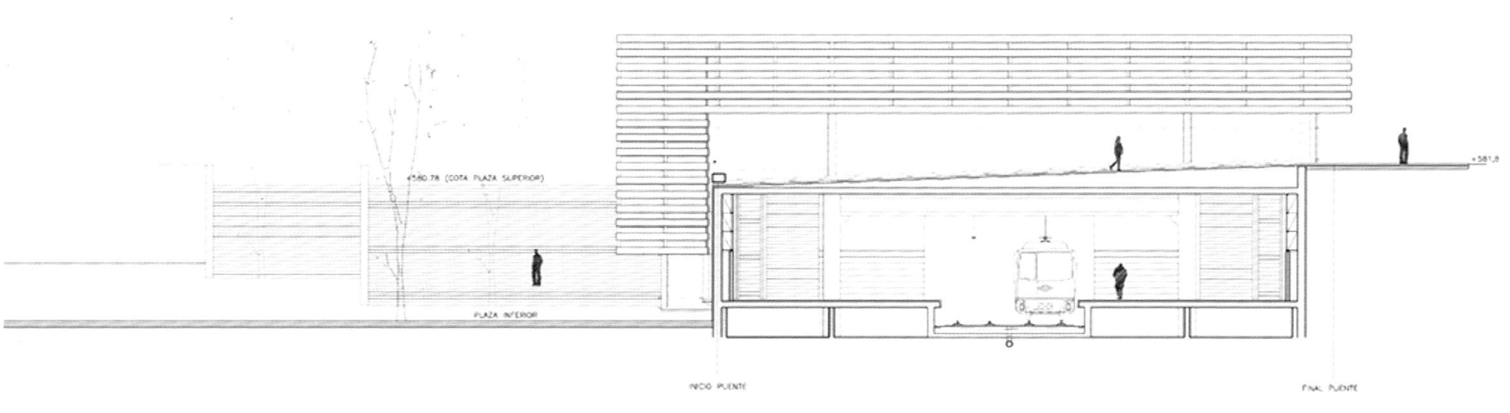

Transverse Section

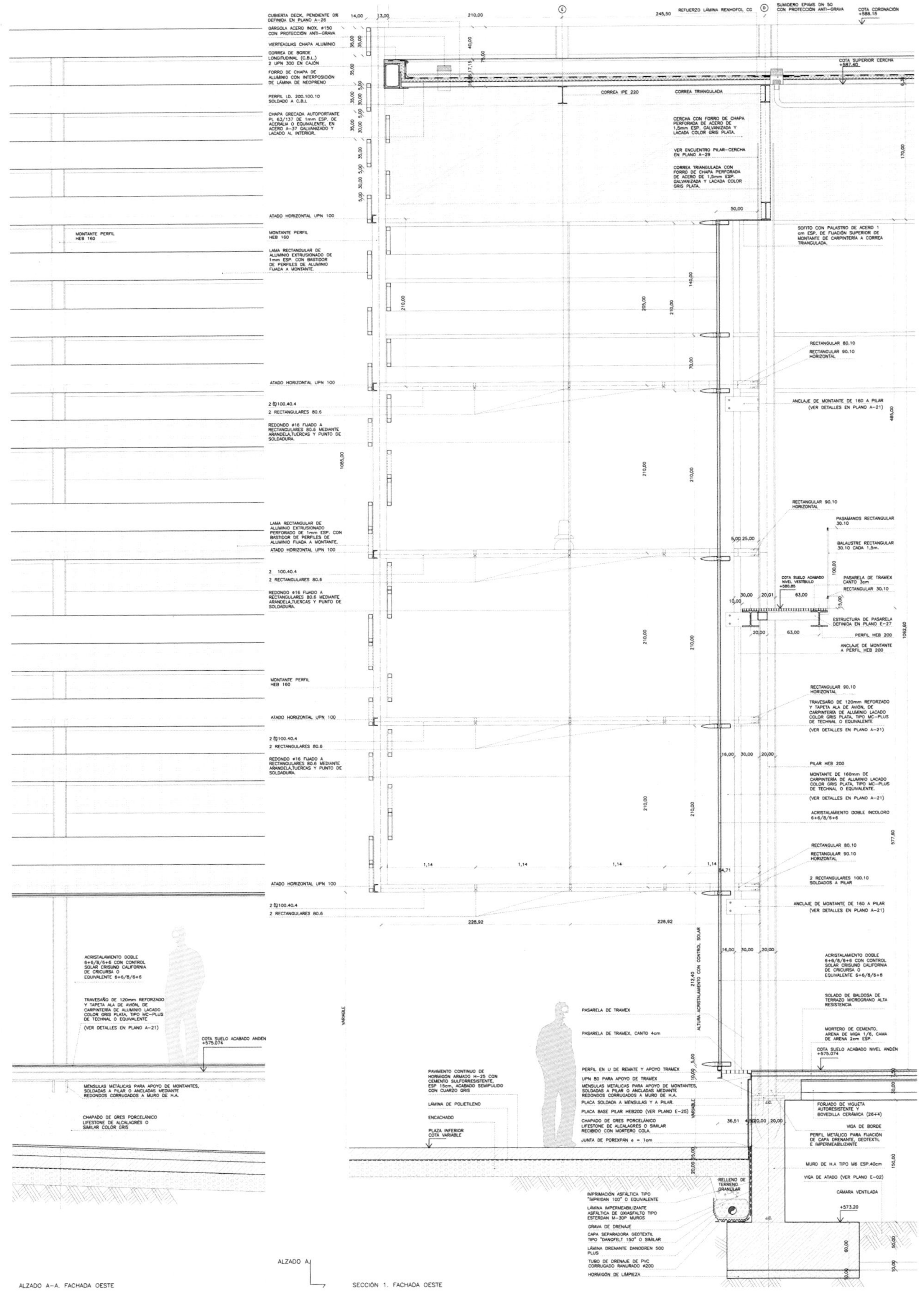

Section Detail

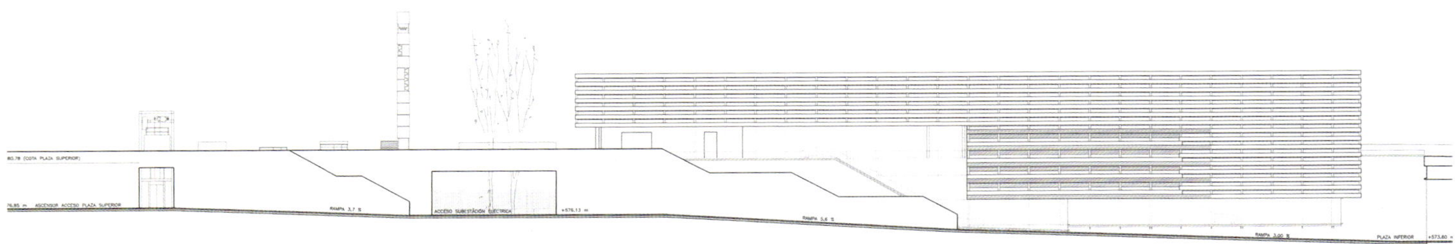

Longitudinal Elevation

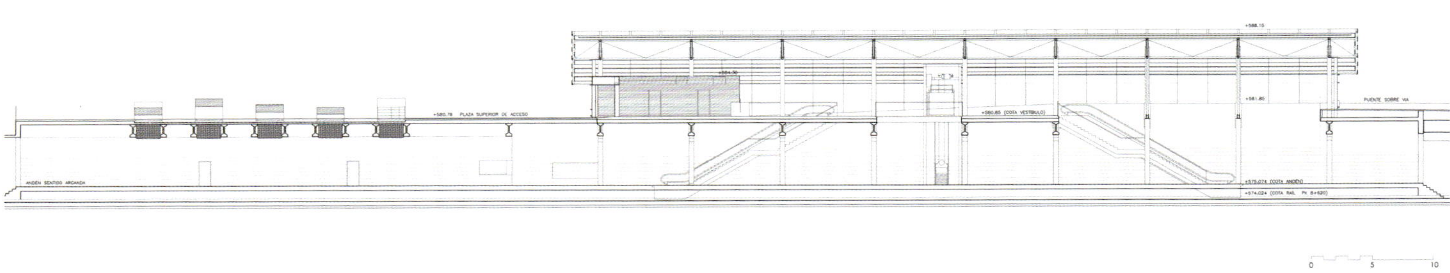

Longitudinal Section

© Miguel de Guzmán

© Miguel de Guzmán

Science Island Museum, Lithuania

SMAR Architecture Studio

1. Reflective Surface.
2. LED Surface,

Natural Light

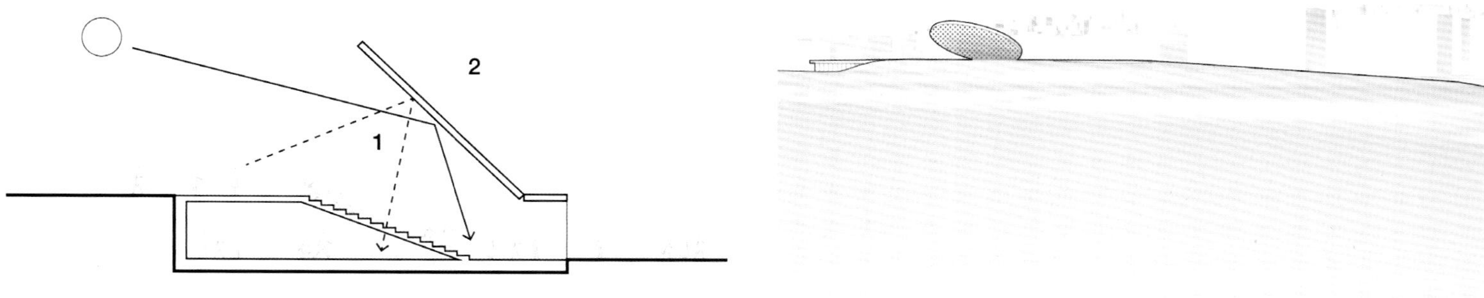

Project Diagram

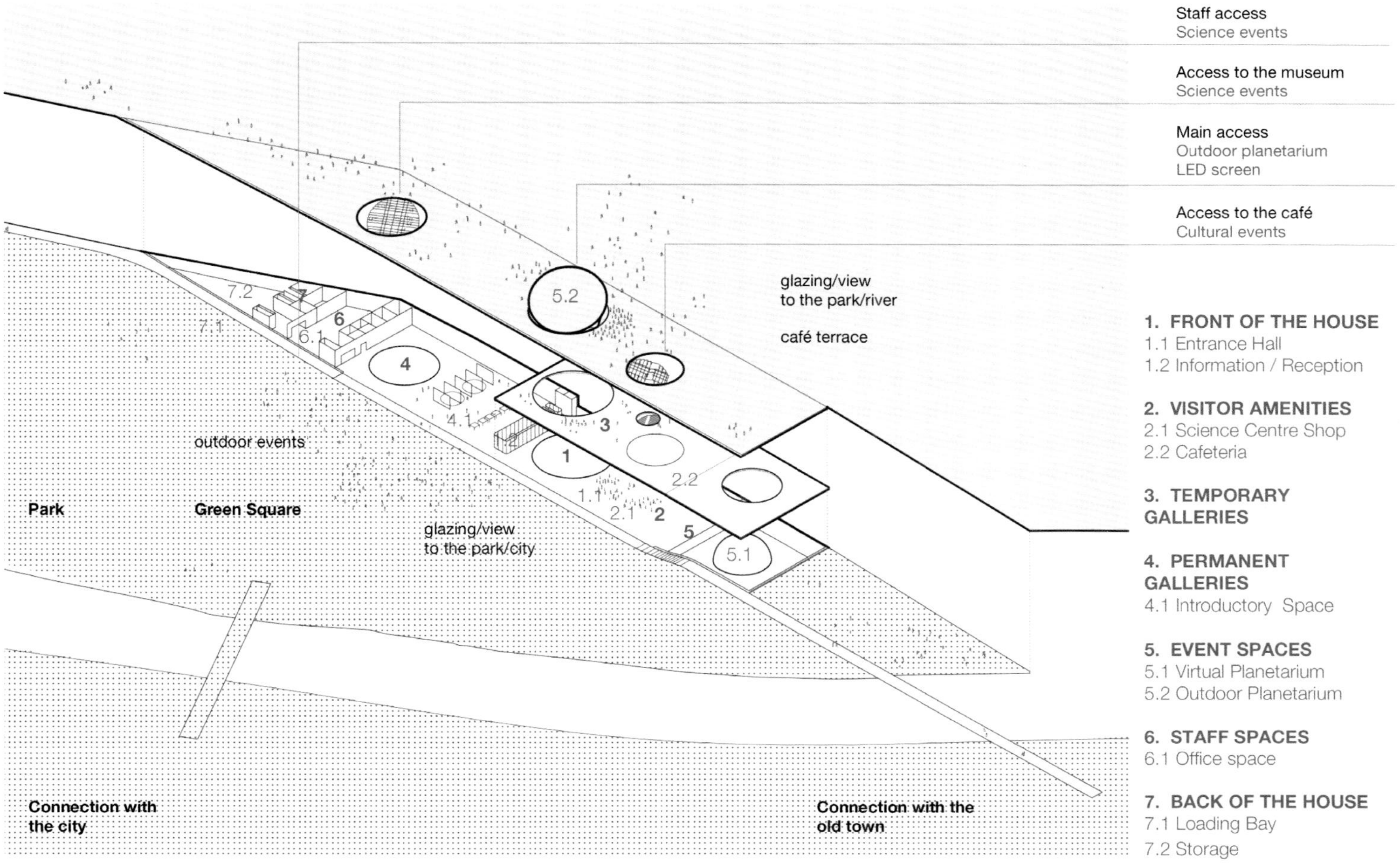

Staff access
Science events

Access to the museum
Science events

Main access
Outdoor planetarium
LED screen

Access to the café
Cultural events

glazing/view
to the park/river

café terrace

7.2 7 6 7.1 6.1 4 4.1 3 5.2 1 1.1 2.1 2 2.2 5 5.1

outdoor events

Park

Green Square

glazing/view
to the park/city

Connection with the city

Connection with the old town

1. FRONT OF THE HOUSE
1.1 Entrance Hall
1.2 Information / Reception

2. VISITOR AMENITIES
2.1 Science Centre Shop
2.2 Cafeteria

3. TEMPORARY GALLERIES

4. PERMANENT GALLERIES
4.1 Introductory Space

5. EVENT SPACES
5.1 Virtual Planetarium
5.2 Outdoor Planetarium

6. STAFF SPACES
6.1 Office space

7. BACK OF THE HOUSE
7.1 Loading Bay
7.2 Storage

Program Diagram

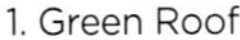

1. Green Roof
2. First Floor
3. Ground Floor
4. Facade
4
3
2
1

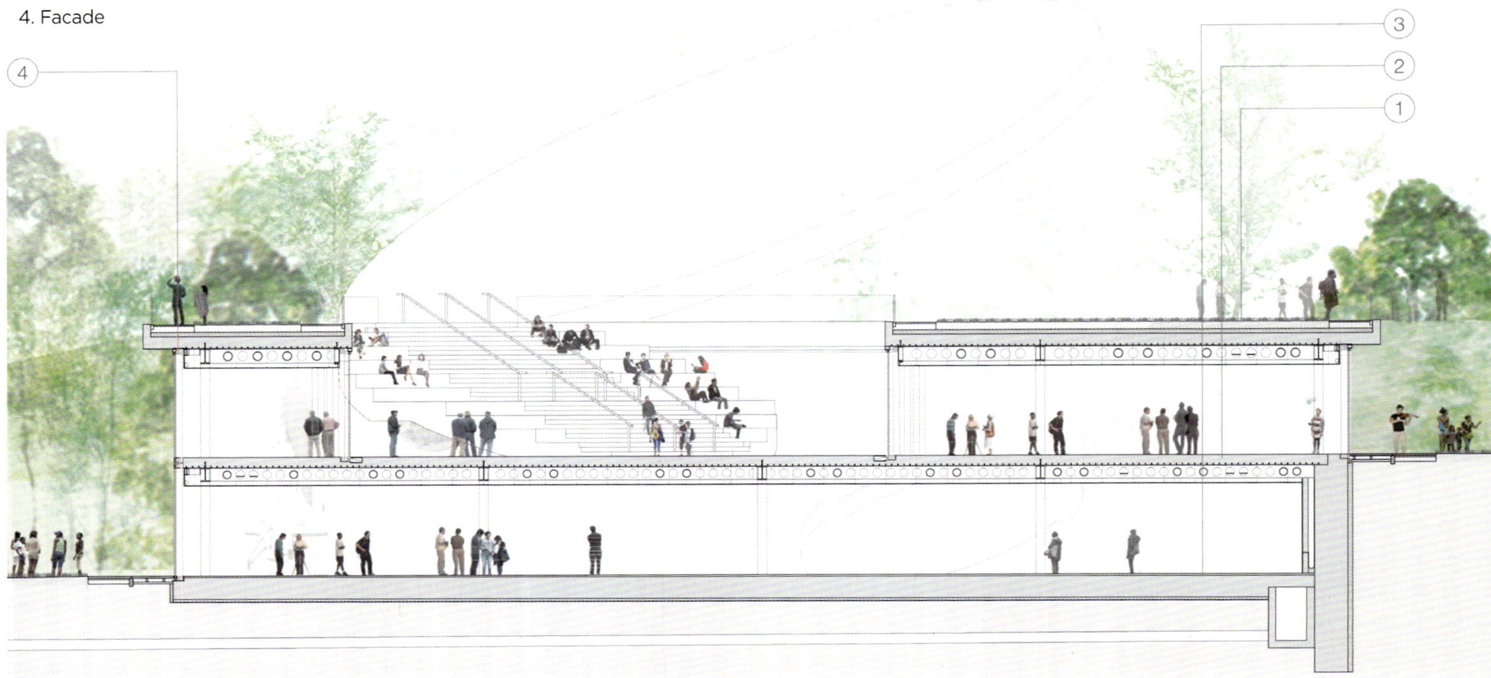

Section

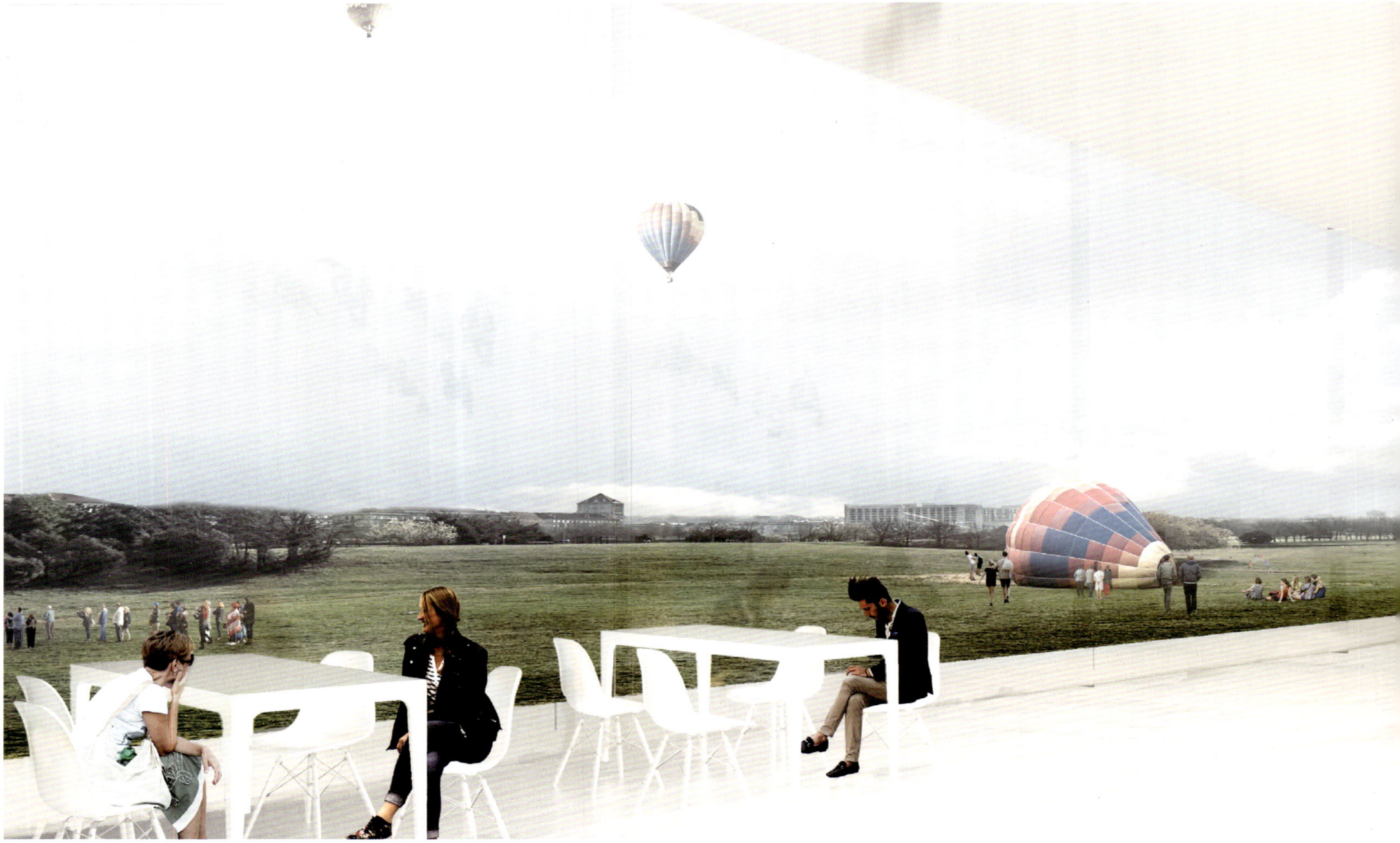

Toll Stations, A16 Motorway, France

Manuelle Gautrand

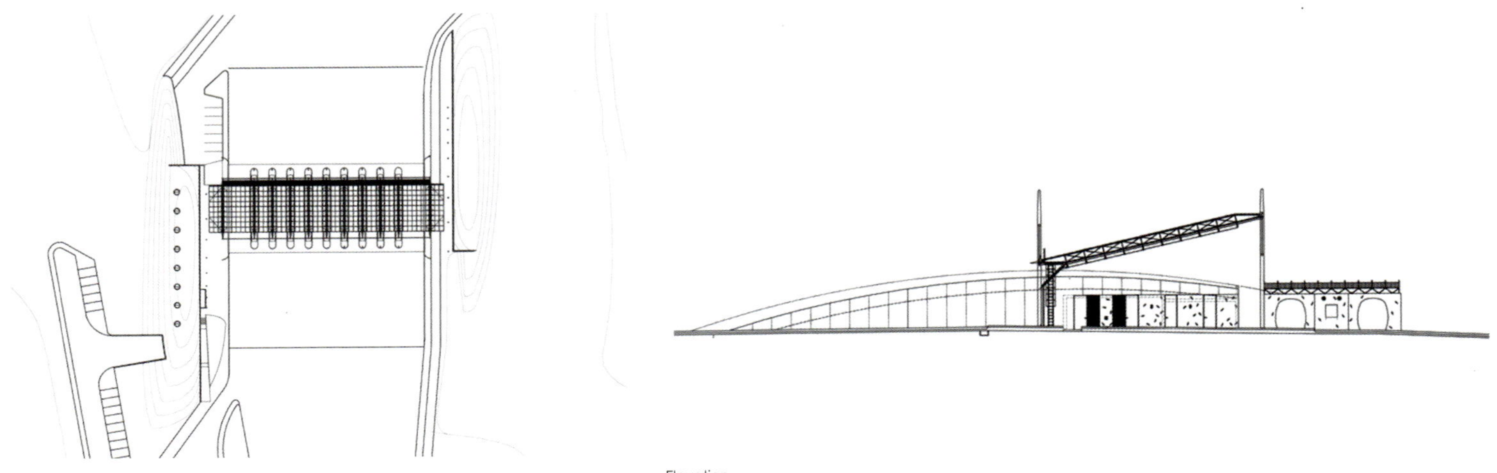
Site Plan

Elevation

Architectural Material Series

To be Continued

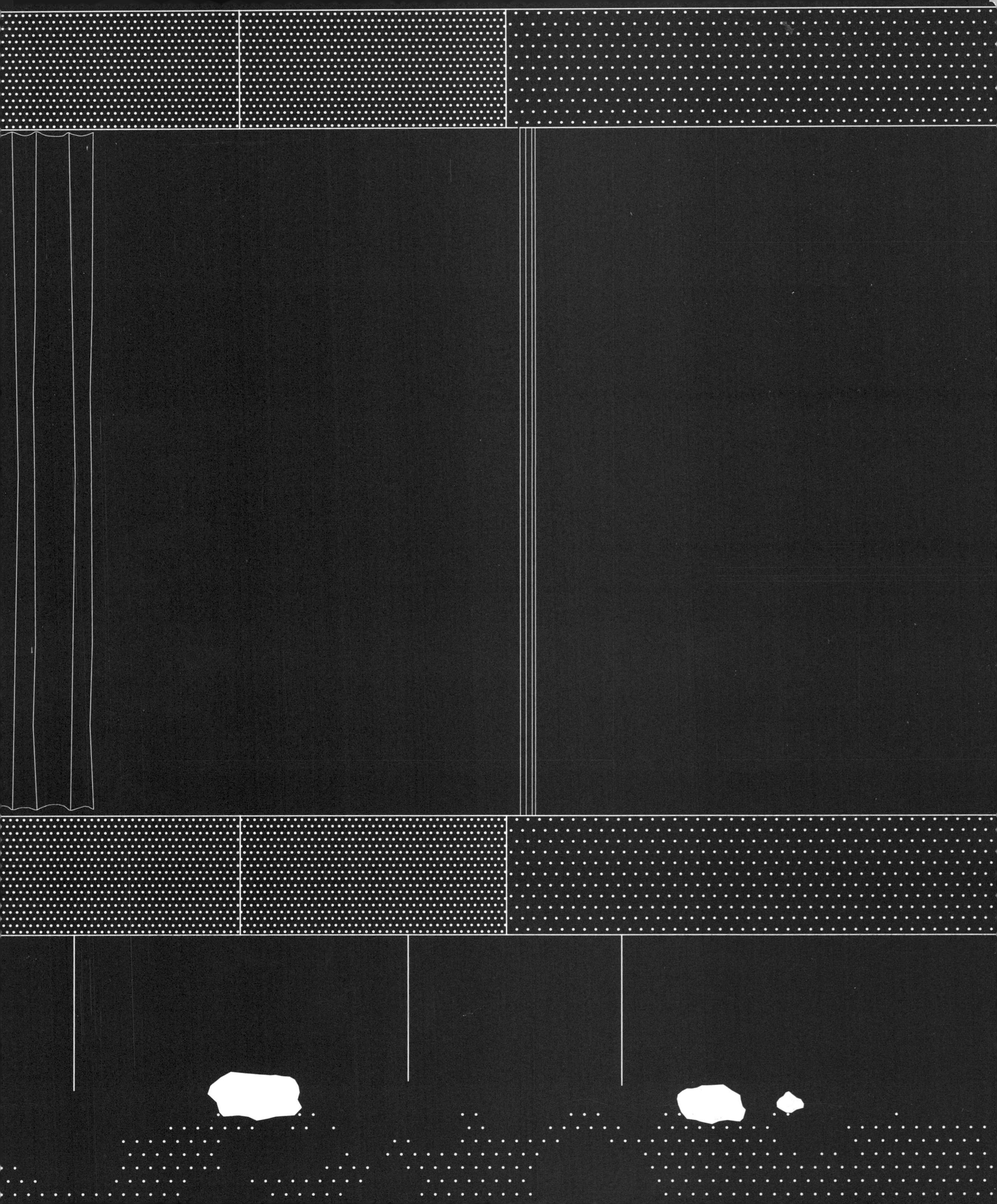